KB010074

July 1st ▸ 27th, 2022

Carl & June's Alaskan Road Trip

Carl과 June의 Alaska 자동차 여행일기

Murrieta, California → Homer, Alaska   7,986 miles (12,777 km) Over 27 days

77 years young Korean Carl, Kwangyong Ahn

77세 한국 젊은이 안광용 저

Completing
the Next-to-Last Item
On My Bucket List
나의 버킷리스트
마지막에서
두 번째 것을 완성!

JINMYONG PUBLISHERS, INC.
www.jinmyong.com

GREENLAND
Godthab
Reykjavik
ICELAND
Arctic
CANADA
Baker Lake
Rankin Inlet
Churchill
Fort McMurray
Thompson
Flin Flon
Prince Albert
Saskatoon
Calgary
Swift Current
Brandon
Williston
Grand Forks
Butte
Inukjuak
Labrador City
Moosonee
Timmins
Amos
Ottawa
Montreal
Duluth
Minneapolis
Milwaukee
Chicago
Detroit
Buffalo
Boston
New York
Philadelphia
Baltimore
Washington D. C.
Scottsbluff
Omaha
Denver
Kansas City
St. Louis
Provo
U. S. A.
Memphis
Norfolk
Charlotte
Albuquerque
Tucson
Birmingham
Atlanta
Dallas
New Orleans
Jacksonville
Houston
Tampa
ATLANTIC OCEAN
Bermuda
Dublin
IRELAND
Cork
Porto
PORTUGAL
Lisbon
Açores
Gibraltar
Madeira
Casablanca
MOROCCO
Canary Islands
MEXICO
Mexico City
Havana
THE BAHAMAS
CUBA
JAMAICA
DOM. REP.
HAITI
Tropic of Cancer
WESTERN SAHARA (occupied by Morocco)
MAURITANIA
Nouakchott
CABO VERDE
Praia
Dakar
SENEGAL
GAMBIA
GUINEA BISSAU
GUINEA
Conakry
Freetown
SIERRA LEONE
Monrovia
LIBERIA
BELIZE
GUATEMALA
HONDURAS
EL SALVADOR
NICARAGUA
COSTA RICA
PANAMA
Caracas
VENEZUELA
GUYANA
SURINAME
FRENCH GUIANA
Bogota
COLOMBIA
Equator
Quito
ECUADOR
Galapagos
Talara
Manaus
Belém
Fortaleza
Teresina
Recife
Trujillo
Porto Velho
Cachimbo
BRASIL
Lima
PERU
BOLIVIA
Barreiras
Aracaju
Salvador
Brasilia
Cuiaba
La Paz
Sucre
Porto Seguro
GoiÂnia
Belo Horizonte
PARAGUAY
Campinas
Tropic of Capricorn
Rio de Janeiro
São Paulo
Curitiba
Antofagasta
Easter Island
CHILE
Asuncion
Porto Alegre
ARGENTINA
Santiago
Rosario
Buenos Aires
Montevideo
URUGUAY
Concepcion
Valdivia
Neuquen
Comodoro Rivadavia
FALKLAND ISLANDS
Stanley
SOUTH GEORGIA ISLAND
1000 Km
Scale at Equator

OCEAN
FINLAND
Alta
Murmansk
Kiruna
Lulea
Skelleftea
Ostersund
SWEDEN
Vaasa
Archangelsk
Vorkuta
Antipayuta
Volochanka
Norilsk
Igarka
Khatanga
RUSSIA
Mirnyj
Syktyvkar
Nizhnevartovsk
Oslo
Stockholm
Helsinki
St. Petersburg
ESTONIA
Novgorod
Vologda
Jaroslavl
Kirov
Perm
Rostov
LATVIA
LITHUANIA
Copenhagen
Minsk
Moscow
Gorkiy
Kazan
Izevsk
Ufa
Sverdlovsk
Chelyabinsk
Omsk
Novosibirsk
Krasnojarsk
Bratsk
Irkutsk
Ulan Ude
Berlin
POLAND
Warsaw
BYELARUS
Orel
Voronez
Tol Yatti
Kuybyshev
Rudnyy
Astana
GERMANY
Krakow
Lviv
Kiev
UKRAINE
Saratov
Aktyubinsk
KAZAKHSTAN
Qaraghandy
Hovd
Ulaanbaatar
Altay
MONGOLIA
CZECH
MOLDOVA
Volgograd
ROMANIA
Bucharest
Frunze
Aralsk
Balqash
Karamay
Urumqi
Dalandzadgad
RUSSIA
Krasnodar
Constanta
BULGARIA
ITALY
Rome
Napoli
UZBEKISTAN
Bishkek
Alma Ata
Korla
Hami
KYRGYZSTAN
Tashkent
Kashi
TURKMENISTAN
Ashgabat
TAJIKISTAN
Baotou
Beijing
Yumen
Izmir
Ankara
TURKEY
Adana
Athens
Tunis
MALTA
TUNISIA
Tripoli
Tehran
Bakhtaran
Esfahan
Qiemo
Hotan
Golmud
Xining
Yinchuan
Taiyuan
Lanzhou
Kaifeng
Xian
Kabul
AFGHANISTAN
Islamabad
CHINA
Hefei
CYPRUS
LEBANON
ISRAEL
SYRIA
IRAQ
Baghdad
JORDAN
Basra
IRAN
Shiraz
Multan
Mianyang
Chengdu
Cairo
KUWAIT
New Delhi
NEPAL
Kathmandu
BHUTAN
Zigong
Yueyang
Guiyang
Sabhah
EGYPT
BAHRAIN
QATAR
PAKISTAN
Jaipur
Agra
Allahabad
(Calcutta)
Kolkata
BANGLADESH
Dhaka
Dukou
Djanet
LIBYA
Al Jawf
Aswan
Riyadh
U. A. E.
Muscat
Ahmadabad
Nanning
Shantou
Mecca
SAUDI ARABIA
Surat
Nagpur
MYANMAR
Hanoi
LAOS
Hong
Bilma
NIGER
CHAD
SUDAN
Khartoum
ERITREA
YEMEN
OMAN
(Bombay) Mumbai
Pune
Rangoon
Vientiane
THAILAND
Sanaa
INDIA
Bangkok
CAMBODIA
VIETNAM
N'Djamena
DJIBOUTI
SOMALIA
Chennai (Madras)
Phnom Penh
Abuja
CENTRAL AFRICAN REPUBLIC
Waw
Addis Ababa
ETHIOPIA
Cochin
SRI LANKA
Ho Chi Minh City
CAMEROON
Yaounde
S.SUDAN
Colombo
MALAYSIA
BRUNEI
Bangui
UGANDA
Kuala Lumpur
MALAYSIA
SINGAPORE
Libreville
CONGO
KENYA
Mogadishu
MALDIVES
Equator
GABON
DEM. REP. OF CONGO
RWANDA
BURUNDI
Nairobi
Brazzaville
Kinshasa
ANGOLA
TANZANIA
Jakarta
Dar es Salaam
SEYCHELLES
Kacanga
Luanda
Likasi
INDIAN OCEAN
ANGOLA
Huambo
MALAWI
ZAMBIA
MOZAMBIQUE
Lusaka
Harare
Antananarivo
NAMIBIA
ZIMBABWE
MADAGASCAR
Windhoek
Port Hedland
BOTSWANA
Learmonth
Walvis Bay
Gaborone
Tropic of Capricorn
Tolanaro
Pretoria
Maputo
Carnarvon
Luederitz
Bloemfontein
SWAZILAND
LESOTHO
Geraldton
SOUTH AFRICA
Umtata
Perth
Cape Town
Port Elizabeth
Antarctic Circle

Homer
Murrieta
Panama
2차 여행 예정
Beaufort Sea
Baffin Bay
Alaska
Fort Yukon
Fairbanks
Anchorage
Cordova
Gulf of Alaska
Yukon Territory
Whitehorse
Juneau
Northwest Territories
Port Radium
Great Bear L.
Yellowknife
Great Slave L.
Nunavut
Baker Lake
Rankin Inlet
Godthab
Iqaluit
Inukjuak
Hudson Bay
CANADA
Churchill
Manitoba
Fort Chipewyan
Alberta
Fort McMurray
Saskatchewan
British Columbia
Bear Lake
Prince Rupert
Dawson Creek
Grande Prairie
Prince George
Edmonton
Red Deer
Calgary
Kamloops
Vancouver
Victoria
Flin Flon
Prince Albert
Saskatoon
Regina
Swift Current
Thompson
L. Winnipeg
Brandon
Winnipeg
Ontario
Moosonee
Timmins
Quebec
Amos
Labrador City
Ottawa
Montreal
Augusta
Maine
Fredericton
Newfoundland
PACIFIC OCEAN
Washington
Seattle
Spokane
Portland
Oregon
Idaho
Boise
Twin Falls
Montana
Helena
Butte
Williston
Grand Forks
North Dakota
Bismarck
South Dakota
Minnesota
Duluth
Lake Superior
Minneapolis
St. Paul
Wisconsin
Milwaukee
Lake Michigan
Lake Huron
Lake Ontario
Lake Erie
Toronto
Rochester
Buffalo
Detroit
Michigan
Chicago
Illinois
Indiana
Indianapolis
Ohio
Cleveland
Pittsburgh
Pennsylvania
Dayton
Cincinnati
West Virginia
Virginia
Richmond
Norfolk
Washington D. C.
Baltimore
Philadelphia
New York
Hartford
Providence
Boston
Massachusetts
Concord
Wyoming
Cheyenne
Scottsbluff
Nebraska
Omaha
Iowa
Salt Lake City
Provo
Utah
Nevada
Reno
San Francisco
Oakland
Las Vegas
California
Los Angeles
San Diego
Mexicali
Tucson
Boulder
Denver
Colorado
Kansas
Kansas City
U. S. A.
St. Louis
Missouri
Kentucky
Louisville
Nashville
Tennessee
North Carolina
Charlotte
South Carolina
Columbia
Santa Fe
New Mexico
Oklahoma City
Oklahoma
Little Rock
Arkansas
Memphis
Birmingham
Mississippi
Alabama
Atlanta
Georgia
Dallas
Texas
Austin
San Antonio
Houston
Jackson
Baton Rouge
Louisiana
New Orleans
Tallahassee
Jacksonville
Florida
Orlando
Tampa
West Palm Beach
Miami
Nassau
ATLANTIC OCEAN
Gulf of Mexico
Hermosillo
Chihuahua
MEXICO
Tropic of Cancer
Monterrey
Culiacan
La Paz
Durango
Ciudad Victoria
Zacatecas
San Luis Potosi
Tepic
Leon
Guadalajara
Queretaro
Morelia
Colima
Toluca
Mexico City
Chilpancingo
Oaxaca
Villahermosa
Merida
Campeche
Chetumal
Belize City
BELIZE
Belmopan
HONDURAS
Tegucigalpa
Guatemala
GUATEMALA
San Salvador
EL SALVADOR
NICARAGUA
Managua
San Jose
COSTA RICA
Colon
PANAMA
Havana
CUBA
JAMAICA
HAITI
Caribbean Sea
Barranquilla
Medellin
© Geographic Guide
1000 miles
1000 km

Murrieta, California → Homer, Alaska

7,986 miles (Round 12,777km)

Over 27 days

캘리포니아, 뮤리에타 → 알래스카, 호머

7,986 마일 (왕복 12,777km)

27일 간

Murrieta
Panama
2차 여행 예정
CANADA
U. S. A.
MEXICO
Northwest Territories
Yellowknife
Great Slave L.
Baker Lake
Rankin Inlet
Juneau
Bear Lake
British Columbia
Prince Rupert
Dawson Creek
Prince George
Kamloops
Vancouver
Victoria
Fort Chipewyan
Alberta
Fort McMurray
Grande Prairie
Edmonton
Red Deer
Calgary
Saskatchewan
Flin Flon
Prince Albert
Saskatoon
Regina
Swift Current
Brandon
Winnipeg
Manitoba
Churchill
Thompson
L. Winnipeg
Hudson Bay
Inukjuak
Ontario
Moosonee
Timmins
Quebec
Amos
Labrador City
Newfoundland
Prince Edward Island
New Brunswick
Fredericton
Halifa
Ottawa
Montreal
Maine
Augusta
Washington
Seattle
Spokane
Portland
Oregon
Idaho
Boise
Twin Falls
Montana
Helena
Butte
Williston
Grand Forks
North Dakota
Bismarck
Minnesota
Duluth
Lake Superior
South Dakota
Missouri
Minneapolis
St. Paul
Wisconsin
Milwaukee
Lake Michigan
Lake Huron
Lake Ontario
Lake Erie
Michigan
Detroit
Toronto
Rochester
Buffalo
New York
Concord
Boston
Massachusetts
Providence
Hartford
RI
VT
NH
NJ
MD
DE
Wyoming
Cheyenne
Scottsbluff
Nebraska
Omaha
Iowa
Chicago
Illinois
Indiana
Indianapolis
Ohio
Dayton
Cleveland
Pittsburgh
Pennsylvania
Philadelphia
Baltimore
Washington D. C.
West Virginia
Virginia
Richmond
Norfolk
Reno
Nevada
Oakland
Salt Lake City
Provo
Utah
Boulder
Denver
Colorado
Kansas
Kansas City
St. Louis
Cincinnati
Louisville
Kentucky
Las Vegas
California
Angeles
San Diego
Mexicali
Arizona
Phoenix
Tucson
Santa Fe
Albuquerque
New Mexico
Oklahoma City
Oklahoma
Little Rock
Arkansas
Memphis
Mississippi
Nashville
Tennessee
North Carolina
Charlotte
Columbia
South Carolina
Birmingham
Alabama
Atlanta
Georgia
Jackson
Dallas
Texas
Austin
San Antonio
Houston
Baton Rouge
Louisiana
New Orleans
Tallahassee
Jacksonville
Florida
Orlando
Tampa
West Palm Beach
Miami
Rio Grande
ATLANTIC OCEAN
Gulf of Mexico
Nassau
THE BAHAMAS
Havana
CUBA
HAITI
JAMAICA
Caribbean Sea
Hermosillo
Chihuahua
Monterrey
Culiacan
Durango
La Paz
Ciudad Victoria
Zacatecas
San Luis Potosi
Tepic
Leon
Guadalajara
Queretaro
Morelia
Colima
Toluca
Mexico City
Chilpancingo
Oaxaca
Merida
Campeche
Chetumal
Villahermosa
Belize City
BELIZE
Belmopan
HONDURAS
Guatemala
GUATEMALA
Tegucigalpa
San Salvador
EL SALVADOR
NICARAGUA
Managua
San Jose
COSTA RICA
Colon
PANAMA
Barranquilla
Maracaibo
Medellin
© Geographic Guide
0
1000 miles
0
1000 km

To

Angella Kim, *my wife*
Blue Daniel June Loos, *my grandson*
Kim Jin Woo, *my great-grandson*
and
Young Koreans who have dreams

사랑하는 아내 안젤라, 손자 블루, 증손자 진우
그리고 꿈을 가진 한국의 젊은이들에게!

Twenty years from now
You will be more disappointed by the things
You didn't do than by the ones you did do.
So throw off the bowlines.
Sail away from the safe harbor.
Catch the trade winds in your sails.
Explore. Dream. Discover!

— Mark Twain

지금으로부터 20년 후,
당신은 자신이 한 일보다 하지 않은 일들로 인해
더 실망하게 될 것이다.
그러므로 돛줄을 던져라.
안전한 항구를 떠나 항해하라.
무역풍을 돛에 태워라.
탐험하라. 꿈꾸라. 발견하라!

— 마크 트웨인

Since reading Mark Twain's "The Adventures of Tom Sawyer" at a young age, I've loved adventure novels and fishing.

I searched for a year to find a travel partner for this trip, but it was not easy. At 3,993 miles one way, and 7,986 miles (12,777 *km*) round trip, no means an easy distance to travel.
I ended up having to do the trip alone, which worried me at first, but now I feel it added a sense of achievement.

I'd like to thank Park Jooil, who helped my plan, my schedule and accommodation. I also thank my friend, Lee Terry, and the nuns in Temecula Kkotdongnae, who prayed for my safe return as well as many friends of mine in Korea and America who gave me generous encouragement and praise. My thanks also go to my wife and family members, and my travel companion, the most hard-working dog, June (June doesn't like to be driven in cars anymore). Finally I want to express my deep appreciation for all of you.

I recommend this book to every person who would love to go on their own road trip, especially to those who hesitate to venture out.

※English translation was aided by Thomas Frederiksen, the author of Jinmyong Publishing Co.'s "Pure and Simple English" series (12 books).

어려서부터 마크 트웨인의 소설『톰 소여의 모험』등 모험소설과 낚시를 좋아했다.
1년 전부터 여행을 함께할 파트너를 구했으나 쉽지 않았다.
편도가 3,993마일, 왕복 7,986마일 (12,777Km)이니, 결코 만만한 거리가 아니었다. 나는 혼자하기로 결정했고, 처음에는 두려움이 앞섰으나 완주하고 나니 성취감이 훨씬 큼을 느낀다.

일정과 숙박 예약을 도와줬던 캐나다 캘거리의 박주일,
무사귀환을 위해 기도해준 친구 Lee Terry(이은애 회장)와
캘리포니아 테미큘라의 꽃동네 수녀님들,
아낌없는 격려와 칭찬을 보내준 많은 한국과 미국의 친구 친지분들,
나를 지켜준 사랑하는 아내와 가족들,
나의 여행 동반자이자 가장 수고한 반려견 준에게도(준은 이제 자동차 타기를 싫어한다),
모든 분들께 깊은 감사의 마음을 전한다.

자동차 로드 트립(Road Trip)을 좋아하는 분들과
모험을 하기를 망설이는 분들께 추천하고 싶다.

— 안광용 Carl Kwangyong Ahn

010-4425-1012
(USA 951-972-0586)

※국문 교정교열은
캘거리의 김영신의 도움을 받았습니다.

YUKON

Anchorage

Homer

Gulf of Alaska

3,69

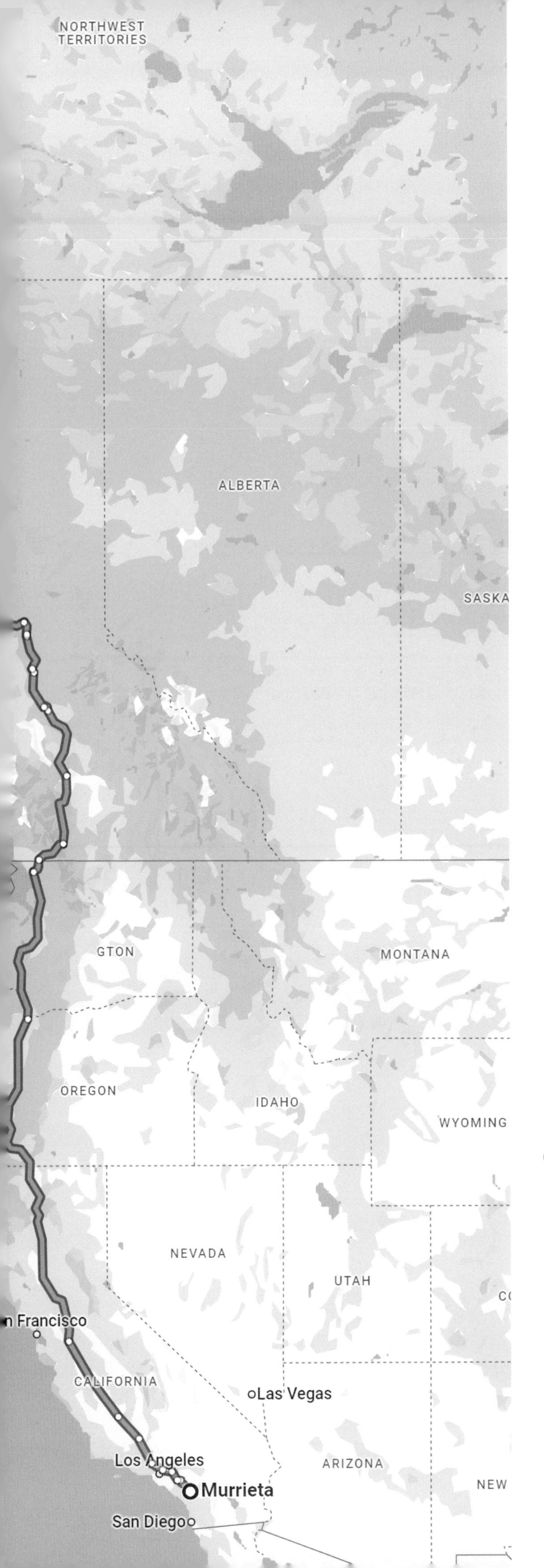

Murrieta, California → Homer, Alaska
7,986 miles (Round 12,777km)
Over 27 days

캘리포니아, 뮤리에타 → 알래스카, 호머
7,986 마일 (왕복 12,777km)
27일 간

Places We Stayed/Itinerary

나와 준이 머무른 곳

1st day

July 1st, Friday

1일째, 7월 1일 금요일

I began the journey at 6:30 a.m.
I left my home in Murrieta, L.A. via the I-15 North and I-91 West Interstate Highways, which are connected to the northbound I-5 Highway that will take me all the way to Canada.
Along the way, I stopped at various rest areas to take breaks from driving, but the rest areas on the American West Coast often don't have any restaurants, just a few vending machines for drinks. After seeing the bill-board for a McDonalds, I exited the highway to grab lunch, before returning to continue my drive north.
I ended up sleeping in Anderson, a small town with a population of around 10,000.
I am still in California.

새벽 6:30 출발.
엘에이, 뮤리에타 집에서 나와 15번 North와 91번 West를 지나 캐나다와 연결되는 5번 고속도로로 달렸다.
중간에 휴게소에 들러서 잠시 쉬기도 했는데 미국 서부 지역의 고속도로 휴게소에는 식당은 없고 음료 자판기, 화장실, 애완견 놀이터 등만 있을 뿐이다. 점심을 먹으러 맥도날드 사인을 보고 고속도로를 벗어났다가 다시 고속도로로 진입하기도 했었다.
인구 10,000여명이 거주하는 앤더슨이라는 작은 도시에 숙박. 아직도 캘리포니아다.

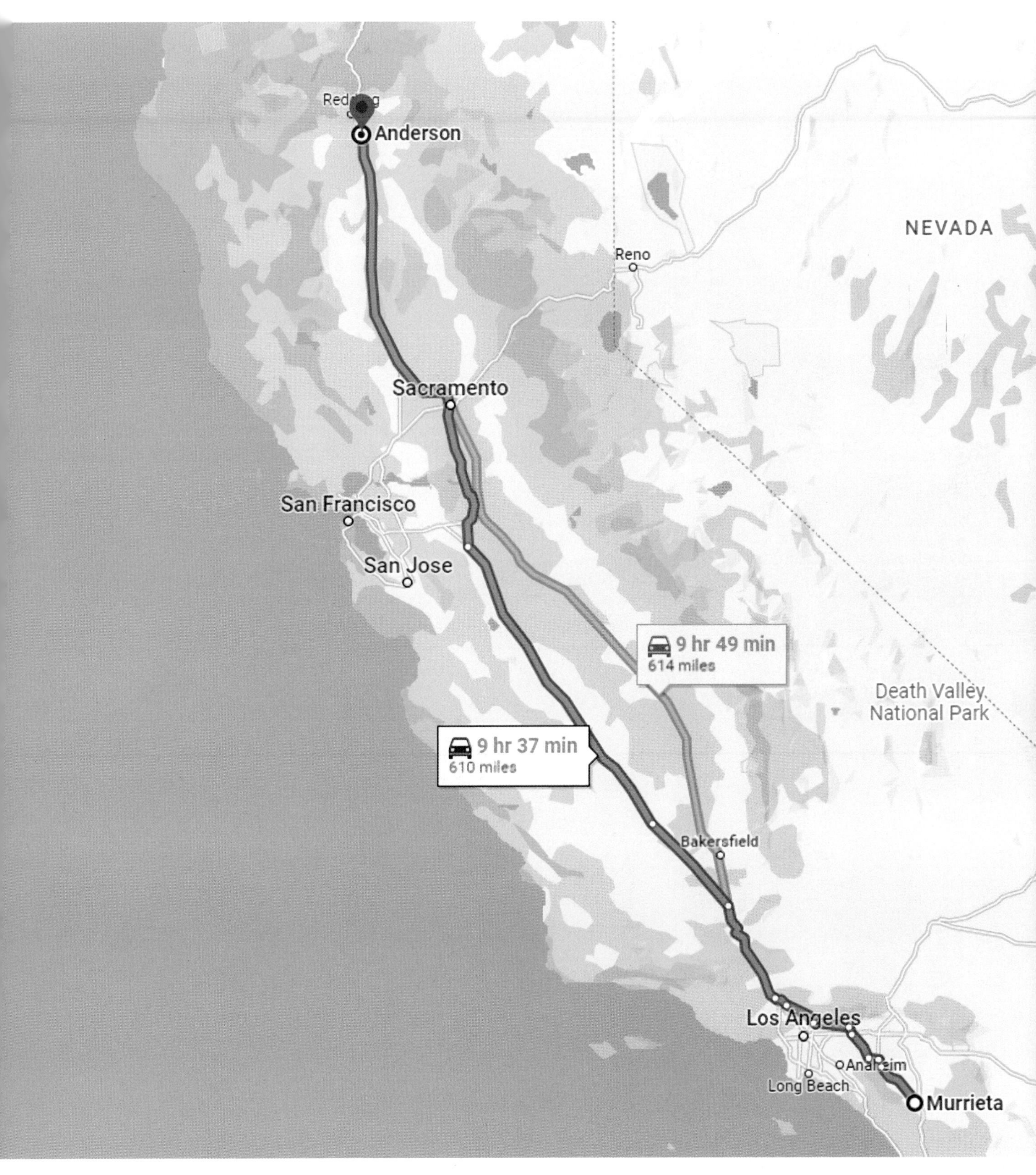

Murrieta → Anderson
610 miles (981 km)

뮤리에타 → 앤더슨, 981 km

ⓘ

June is knocked down after reaching our lodging for the night

숙소에 와서 늘어진 준

A truck carrying a small lightweight plane for personal use

자가용 경비행기를 싣고 가는 차량

2nd day

July 2nd, Saturday

2일째, 7월 2일 토요일

After starting the day off with a light breakfast provided by our motel, we passed all the way through the state of Oregon to reach Seattle, Washington. Originally the plan was to meet my friend Sung Lee for dinner, but I found out he was in quarantine after catching Covid19.
Tomorrow, I will be crossing the border into Canada and staying in Vancouver.

숙소에서 제공하는 간단한 조식으로 아침을 먹고 오레곤주를 지나 미국 워싱턴주 시애틀에 도착. 원래 계획은 친구 이성호를 만나 함께 저녁 먹기로 했는데 코비드 19에 걸려 격리중이라고 한다.
내일은 국경 넘어 캐나다 입국, 밴쿠버에서 잘 생각이다.

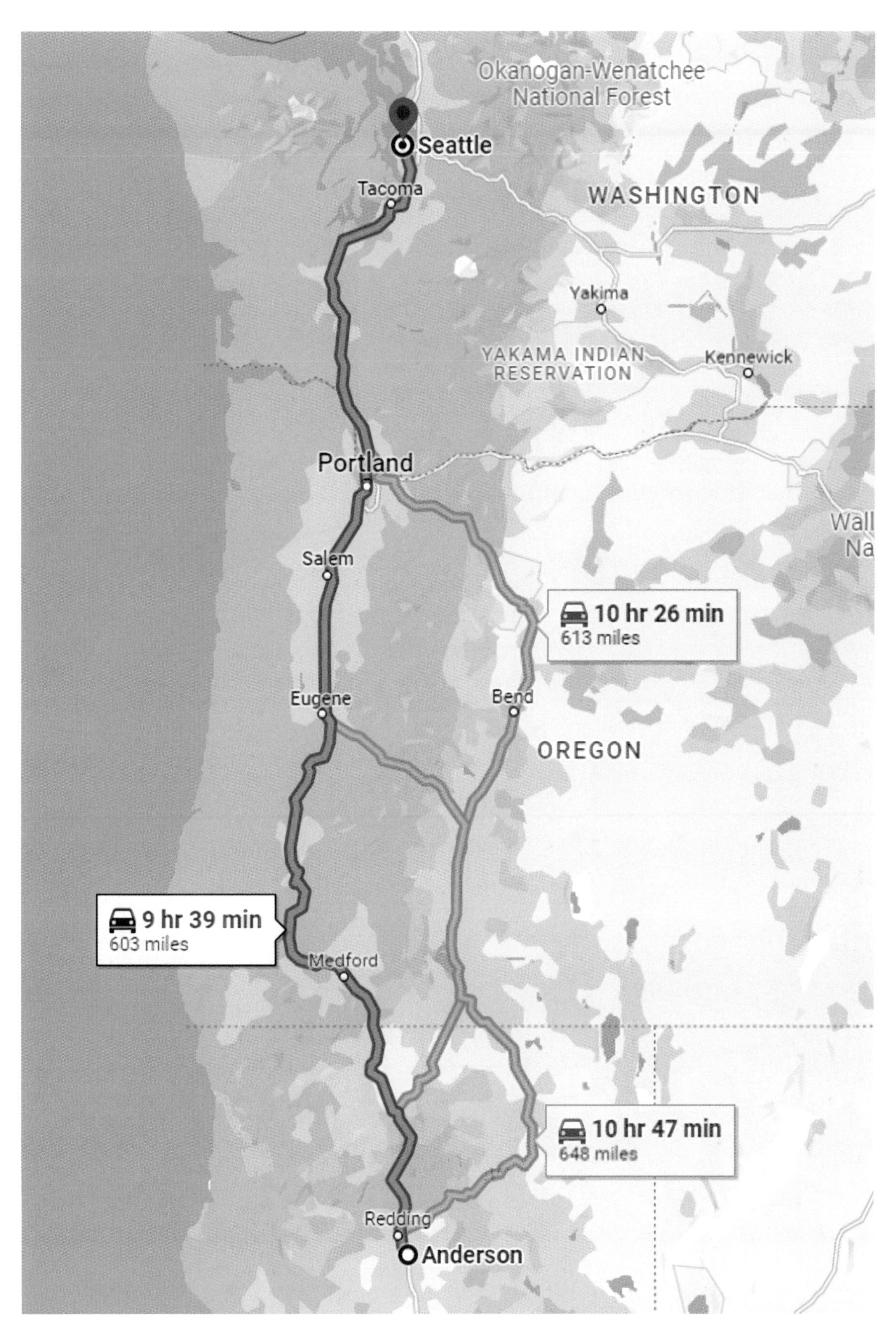

Anderson → Seattle
603 miles (970 km)

뮤리에타 → 시애틀, 970 km

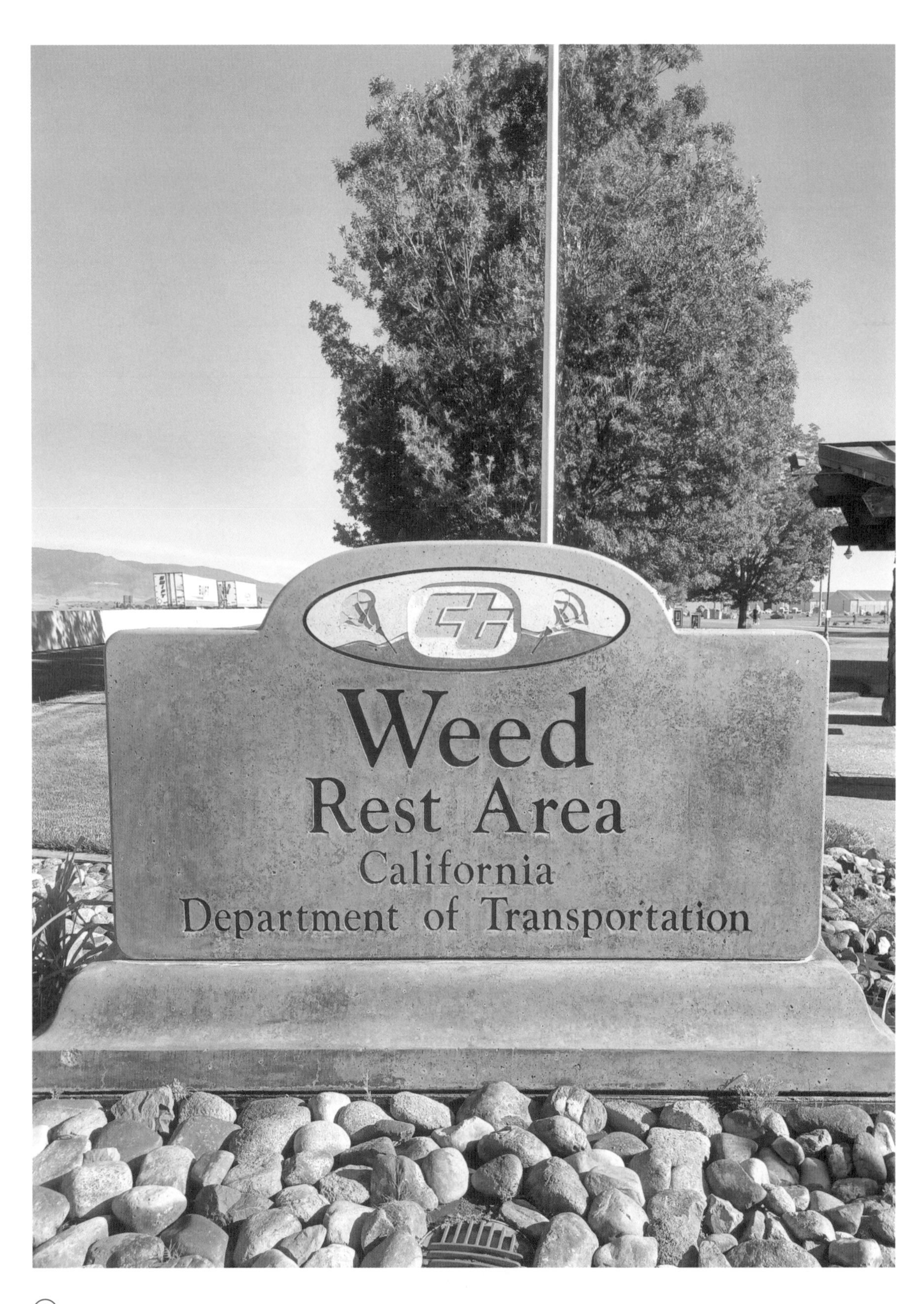

The Weed Rest Area.

캘리포니아 위드 휴게소

↑

A snow-covered mountaintop visible from the Weed Rest Area.

위드 휴게소에서 바라본 정상에 눈이 덮인 산

3rd day

July 3rd Sunday

3일째, 7월 3일 일요일

We crossed the border into Vancouver, Canada.

I was slightly worried about the Covid19 test I would have to take (not to mention the oranges and figs that I'd gathered from my Murrieta home to give to my niece, which I was now trying to smuggle in), but we crossed the border safely.

Vancouver is the home of my sister's youngest daughter, Yoon-hee. We met up with my niece and had a hearty dinner of bulgogi and soybean paste stew at a Korean restaurant, together with her two girls Yuri and Mari, and their dog, Tori, before moving to their place for some fruit dessert, and an early rest at that night's accommodation.

국경 넘어 캐나다, 밴쿠버 도착.

코비드19 검사와 차에 실린 오렌지와 무화과(집에서 따온 것으로 조카선물)로 걱정은 좀 했는데 국경 무사히 통과.

밴쿠버는 누나의 작은 딸인 윤희가 사는 곳이다. 조카 손녀인 유리, 마리, 그리고 토리(개)와 함께 한식 레스토랑에서 불고기백반, 된장찌개로 푸짐하게 저녁을 먹고, 윤희네 집으로 와서는 디저트로 과일을 먹은 후 숙소로 돌아와 일찍 쉬었다.

Seattle → Vancouver
143miles (230 km)

시애틀 → 밴쿠버, 230 km

A procession of cars waiting
bumper to bumper
to cross the Canadian border.

꼬리에 꼬리를 문 국경을 넘는 차량 행렬들

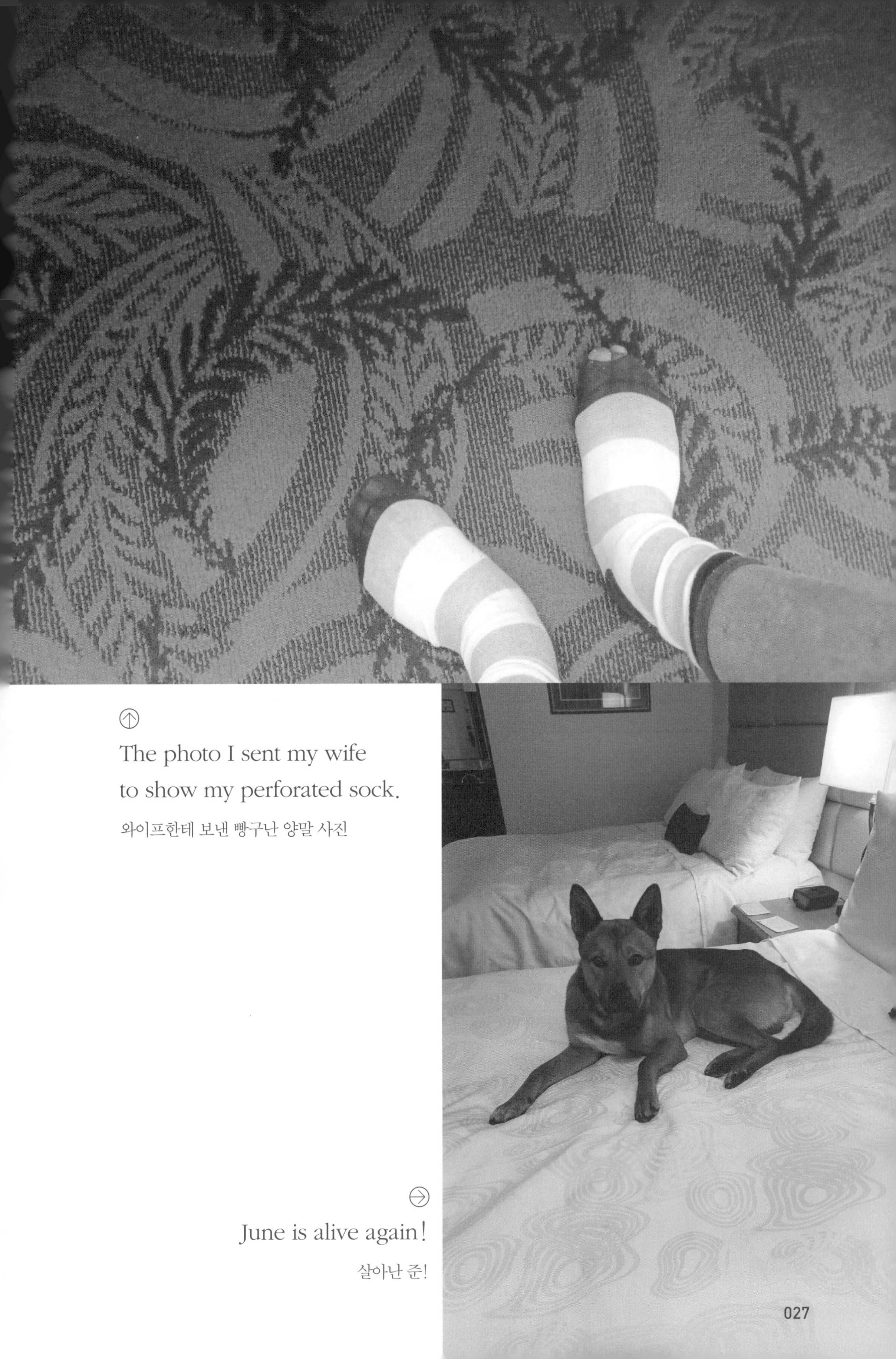

↑

The photo I sent my wife
to show my perforated sock.

와이프한테 보낸 빵구난 양말 사진

→

June is alive again!

살아난 준!

4th day

July 4th, Monday

4일째, 7월 4일 월요일

Around 9 A.M.

I had breakfast with Yoon-Hee and her family at IHOP (International House of Pancakes), and got them to wrap my leftovers, which would take care of lunch as well.

Around 6 P.M., we had dinner at a small Chinese buffet restaurant.

I'd reserved a small motel for the night around 60 miles from here.

We checked in to my motel in the town of 100 Mile House around 8:30 P.M.

오전 9시경.

나의 숙소 가까이 있는 IHOP(International house of pancake) 팬케익 전문 식당에서 윤희 가족과 함께 조식을 먹었다. IHOP에서 싸 온 남은 음식으로 점심까지 해결.

오후 6시경 조그만 차이니스 뷔페에서 저녁식사.

여기서 60마일 더 가면 예약한 숙소가 있다.

오후 8시 30분.

100마일 하우스라는 작은 동네 한 모텔에 숙박.

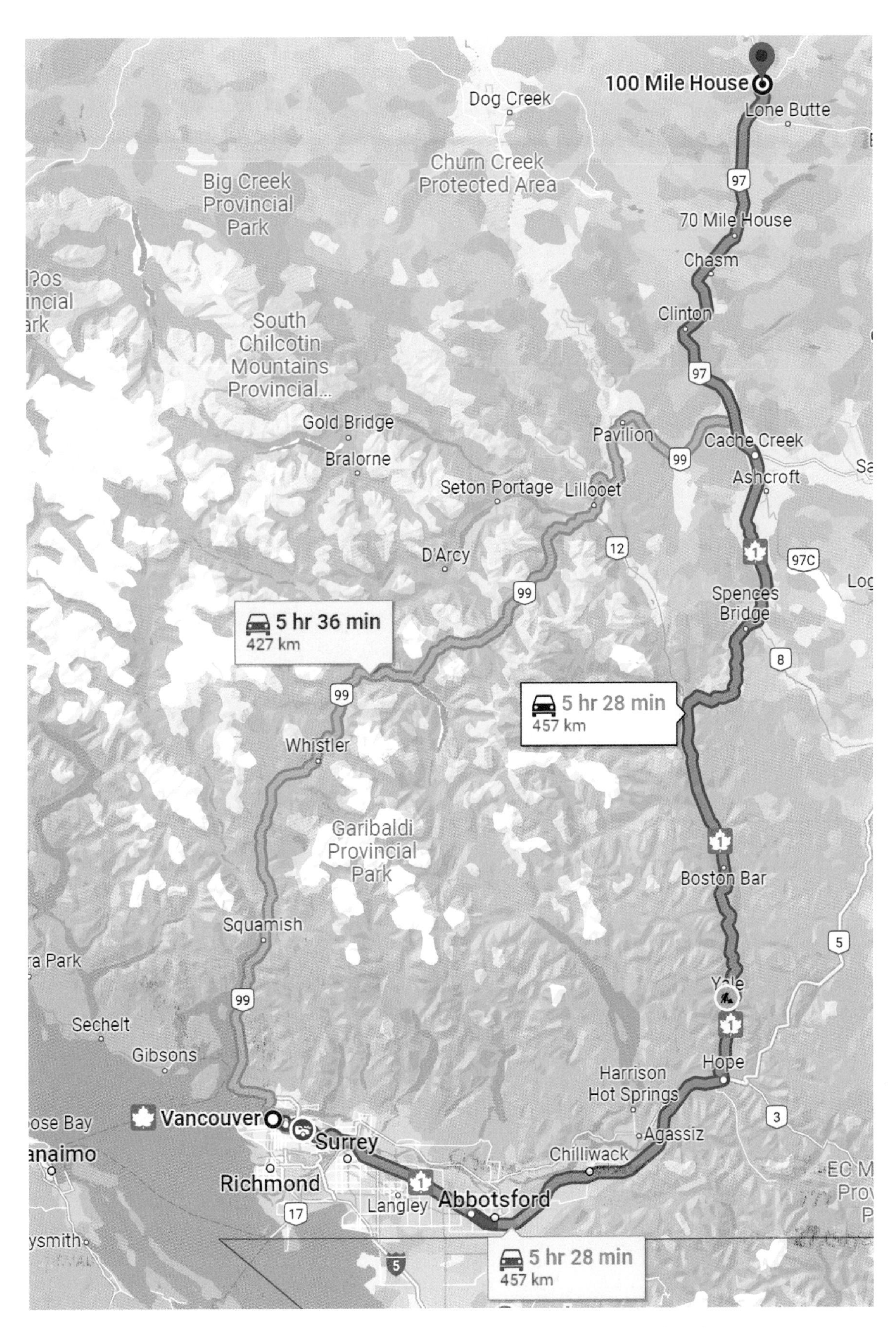

Vancouver → 100 Mile House
283 miles (457 km)

밴쿠버 → 100마일하우스, 457 km

Ⓘ

Break time in 100 Mile House

모텔 앞에서 기념 촬영

This Denny's restaurant took care of breakfast for me.

아침 식사를 해결한 Denny's 식당

In a small village,
deep in the Rocky Mountains,
there is this Korean grocery store
run by the Korean couple
Sukhee and Mr. Jeong.

로키 산맥 깊은 산속 작은 동네에
한국인 Sukhee Jeong 커플이
운영하는 식료품점.

This Chinese restaurant
solved lunch for us.

가는 도중 점심을 해결한 차이니스 레스토랑

5th day

July 5th, Tuesday

5일째, 7월 5일 화요일

We crossed the Rocky Mountains range today.
There are signs here informing us that water and gasoline should be prepared in advance, since they are unavailable upon entering the mountains.
Tonight's accommodation is in Fort St. John, a town of about 20,000 residents.

록키 산맥을 달리고 있다.
여기 진입 전 곳곳에 "물, 휘발유를 미리 준비해서 가야 한다"라고 푯말/안내문이 붙어 있다.
약 20,000 인구의 포트 세인트 존에서 숙박.

100 mile house → Fort St. John
473 miles (762 km)

100마일하우스 → 포트 세인트 존, 762 km

A gas station in the mountains.

산 중에 있는 주유소

Dignified June. He occupied the bed, even though there's only one bed in the room.

의젓한 준. 침대가 하나인데도 먼저 차지함.

6th day

July 6th, Wednesday

6일째, 7월 6일 수요일

It's the 6th day of my adventure, and I'm now 2,124 miles away from home. I fell asleep around 11 P.M. last night, but when I awoke around 5 a.m., it was already bright outside. I guess we're near the North Pole now …
Since we started through the Rocky Mountains for the distance of some 300 kilometers, there was a long stretch of roads with no gas stations or villages at all. It will take three more days till we reach Alaska!
I stopped by a small village in the mountains to look for some place to eat. When I approached a Japanese restaurant, I discovered it had closed down due to Covid19. Fortunately, I found a Chinese restaurant and solved my dinner problem with some wonton soup.
We checked in at the Toad River Lodge. We saw many beautiful lakes. It was truly a magnificent and scenic view.

모험여행 6일째, 집에서 2,124 마일 왔음.
어제 밤 11시 환할 때 잠들었는데 오늘 새벽 5시 눈을 뜨니 이미 밖이 훤하다.
북극점에 가까이 왔나 보다.
어제 록키 산맥 시작되는 지점을 넘었을 때 300km를 조금 넘는 긴 구간 동안 주유소도 마을도 없었다. 3일을 더 가야 드디어 목적지 알래스카!
산속 작은 동네로 들어와 식사할 곳을 찾아 다니다 일식집이 있어 가보니 그곳은 코비드19로 이미 폐점이었다. 다행히 중국식당이 보여 그곳에서 완탕 수프로 저녁을 해결했다. 토드 리버 롯지에 체크인.
아름다운 호수가 많이 보였다. 참으로 광활한 풍경이었다.

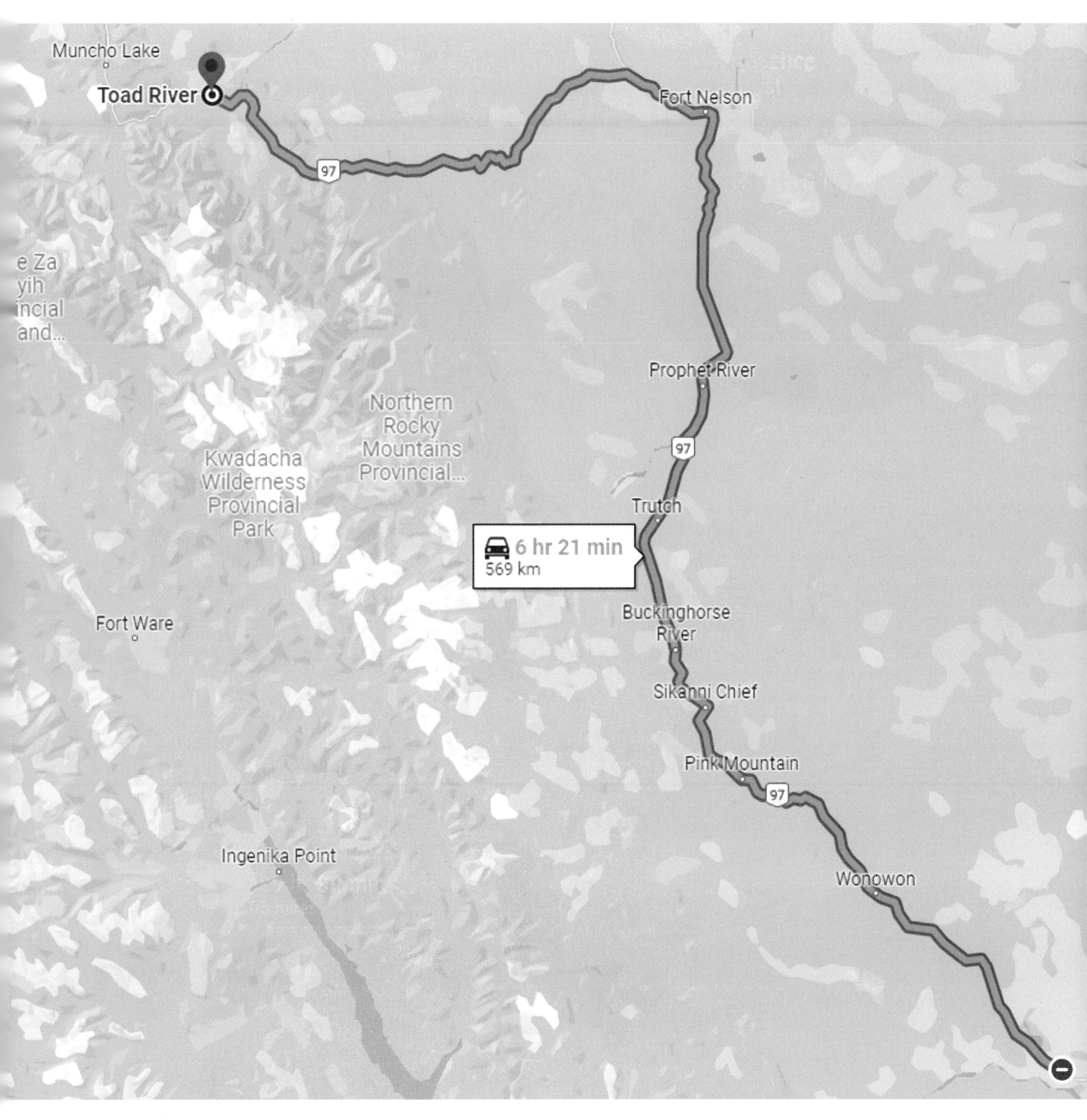

Fort St. John → Toad River
353 miles (569 km)

포트 세인트 존 → 토드 리버, 569 km

There won't be another gas station for 176 kilometers.

176*km* 동안 주유소가 없단다.

A bike-riding traveler whom we met at the highway rest area to cross the Canadian border.

고속도로 휴게소에서 만난 산악자전거 여행객

ⓐⓑⓒ

The scenery at the Toad River Lodge.

토드 리버 롯지 풍경

7th day

July 7th, Thursday

7일째, 7월 7일 목요일

I woke up at 5 a.m. again, and it was already bright outside my window. Did we even have a proper night last night, or are we so far north that we are experiencing the "white night" phenomenon? Today, we're planning to travel to Whitehorse, in the Yukon.

I spent the morning together with Lillianne, a lady I had met the night before. She is a teacher at Alaska's special school for the disabled, and was returning to Alaska after a month and a half spent traveling around the U.S. for the summer. Like me, she was also traveling with her dog, the two of them driving in a Toyota with its own attached summer house (Many travelers here seem to be traveling with dogs, so most motels welcome 4-legged guests as well).

새벽 5시에 눈을 떴는데 아직 창 밖은 훤하다. 백야 현상으로 밤이란 게 있기는 한 건지? 오늘은 유콘 자치주 화이트 호스까지 갈 예정이다.

아침은 어제 저녁에 만난 릴리안과 함께 했다. 그녀는 알래스카 장애인특수학교 교사로 여름방학 동안 미국 전역을 한달 반 여행 후 알래스카로 돌아가는 중이다. 자기 차인 도요타 여름별장을 운전하며 애견과 함께 여행 중이다. 대개의 여행객들이 반려견을 데리고 다니니 모텔에서도 반려견 동반을 환영한다.

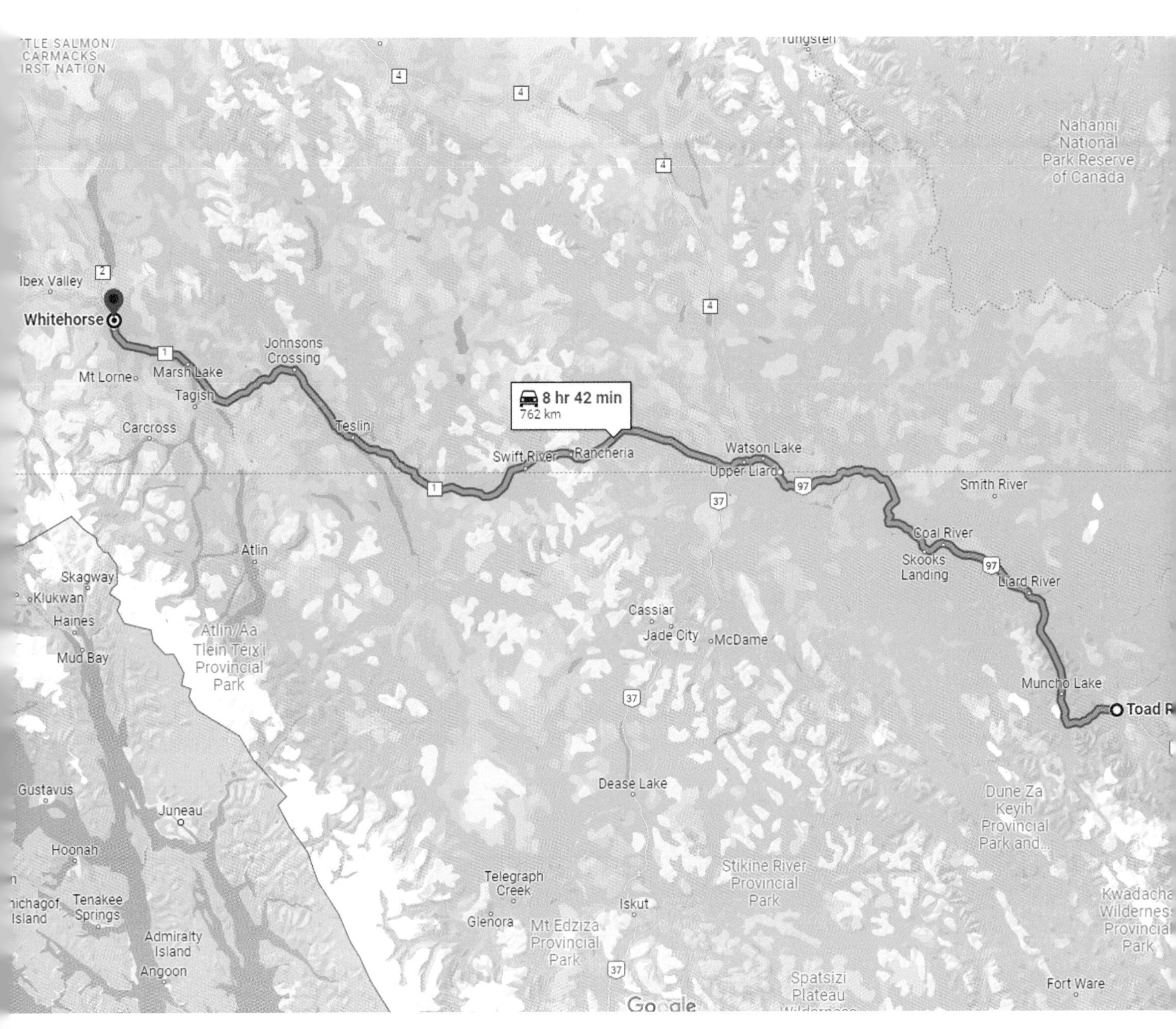

Toad River → Whitehorse
473 miles (762 km)

토드 리버 → 화이트 호스, 762 km

Lillianne, whom I met while traveling.

여행 중 만난 릴리안

Lillianne의 Toyota Summer House

릴리안의 자칭 "여름별장"

ⒶⒷⒸ

Some of the many wild animals we met on the Alaska Hwy.

알래스카 고속도로에서 만난 많은 야생동물들

WHAT IS NO DOMESTIC GARBAGE?

도메스틱 쓰레기란 뭘까?

※편집자 주: 가정용 쓰레기를 말해요.
가정용 쓰레기를 여기 가져와서 버리지 말라는 뜻.

His owner,
his dog,
Two handsome guys!

나의 주인님, 나의 개님!

↑

Some of the motorhomes parked at our lodging.

숙박 중인 모터홈들

→

A gas station at the lodge.

숙소에 있는 주유소

8th day

July 8th, Friday

8일째, 7월 8일 금요일

We crossed the border today and returned back to the U.S.A., this time to our destination of Alaska. In just 7 days, I've managed to reach Alaska, and am now staying in the small riverside city of Tok.
The internet is not working well here, with Wi-Fi and network signals both out of reach.
I feel disconnected from everything in the world …

오늘은 국경을 넘어 다시 미국 알래스카주에 입국. 7일 만에 목적지인 알래스카에 도착해서 톡이란 강가 동네까지 갈 예정이다.
인터넷이 잘 터지지 않는다. 와이파이/네트워크 연결이 안 되는 지역으로 들어와 버렸다.
세상의 모든 것과 단절된 느낌이다.

Whitehorse → Tok
386 miles (622 km)

화이트 호스 → 톡, 622 km

A light-weight seaplane
at a rest area in Alaska.

알래스카의 휴게소에 있는
호수용 경비행기

A quiet rest area.

한산한 휴게소

C-GRZO
SUPER CUB

9th day

July 9th, Saturday

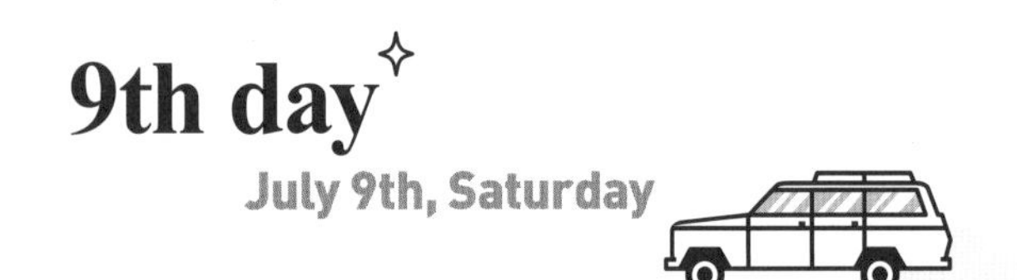

9일째, 7월 9일 토요일

5:30 A.M.

It's now 9 days and 3,350 miles since we left Murrieta. My phone has been dead the entire time since we left the territory of the Yukon. Along the Alaska Highway, you can spot many wild animals, including bears.

We've finally arrived in Alaska's largest city, Anchorage (there's a 4-hour time difference from here to New York and a 1-hour difference from L.A.).

The first thing I did when I arrived was to visit a Korean restaurant, where I ordered soybean paste stew and galbi, as well as rice and kimchi. It's been 5 days since I last ate kimchi, in Vancouver. Finally getting the chance to eat. This kind of Korean food is simply the best and it was the most nutritious energy booster. My travel partner, June, likes the ribs and rice too.

오전 5:30.

뮤리에타 떠난 지 9일째, 3,350마일 지점. 어제 캐나다 유콘 자치주에서부터 하루 종일 전화불통. 알래스카 고속도로에서는 곰을 비롯해 야생동물들을 많이 만나게 된다.

드디어 알래스카에서 제일 큰 도시 앵커리지에 도착(시차로는 뉴욕과 4시간, 엘에이와는 1시간 차이가 난다).

도착해서는 제일 먼저 한국음식점을 찾아가 된장찌개와 갈비를 시켜 밥과 김치랑 먹었다. 5일전 밴쿠버에서 먹고는 김치를 먹지 못했다.

오랜만에 먹은 한국음식은 그야말로 최고의 영양식이고 에너지원이다.

나의 여행 파트너인 준도 갈비와 밥을 좋아한다.

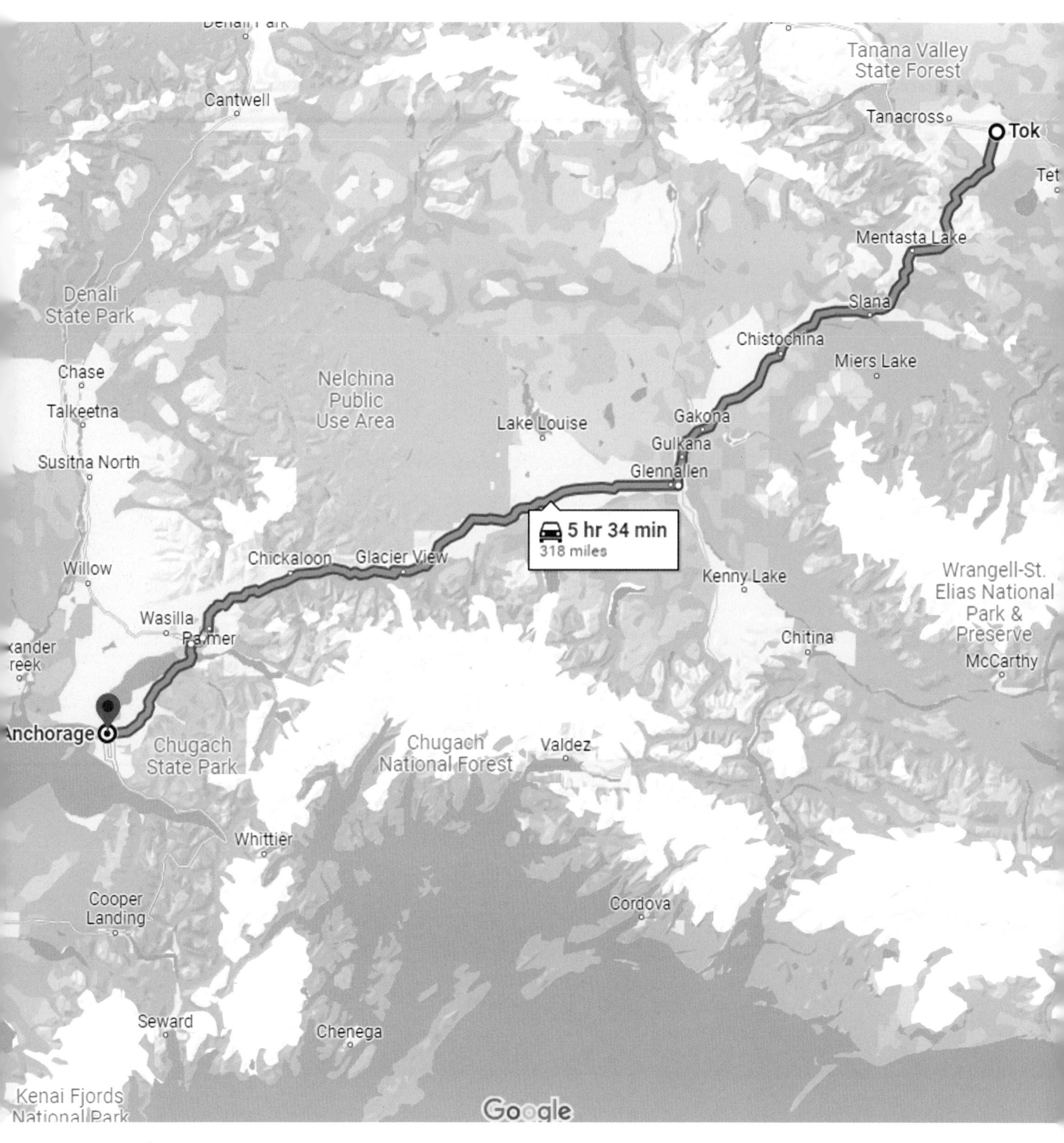

Tok → Anchorage
318 miles (511 km)

톡 → 앵커리지, 511 km

↑

June always lies down to rest as soon as we get in the motel room.

숙소에만 오면 바로 드러눕는 1살 반 준할아버지

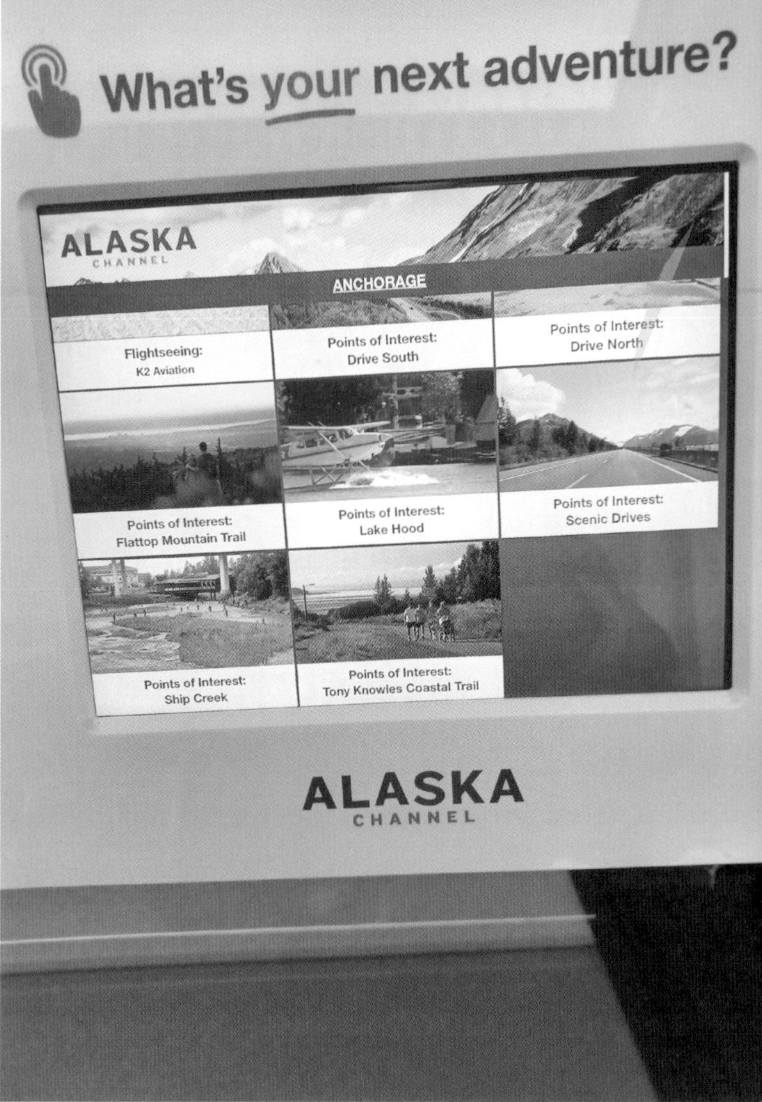

A television channel promoting the points of interest in Alaska.

알래스카를 홍보 소개하는 텔레비전 채널

LA galbi and doenjang jjigae.

갈비와 된장찌개만으로 왕의 밥상!

10th day

July 10th, Sunday

10일째, 7월 10일 일요일

On this our 10th day, I learned that there are no available motels at tonight's destination of Ninilchik.

We had to travel 44 miles to the neighboring city of Soldotna, where we conveniently checked into a motel with a McDonald's right outside it.

Tomorrow, we will arrive at our final destination of Ninilchik (a small fishing village of around 500 people). It's been 20 years since I last was there …

This was a place where my friend Choi Yong-woo and I had been a few times to fish for king salmon and halibut. When I arrive at Ninilchik, I'll try to look up my old friend, Larry and his family. It's been over 20 years since I lost contact with them, so I don't know if they are still living there.

10일째, 오늘 가는 곳 니닐칙에는 모텔이 없다.

44 마일 남겨놓은 지점 솔도트나란 곳에 모텔이 있어 체크인. 모텔 앞에 있는 맥도날드에서 식사 해결.

내일은 이번 모험여행의 최종 목적지인 니닐칙(인구 500여 명의 조그만 어촌)에 도착할 것이다. 20년 만이다.

내 친구 최용우와 함께 대형 연어와 광어 낚시를 몇 번 왔던 곳이다.

니닐칙에 가면 옛 친구 래리의 가족을 찾아볼 생각이다. 연락 두절된 지 벌써 20여년이 지났다. 지금도 그곳에 살고 있는지 궁금하다.

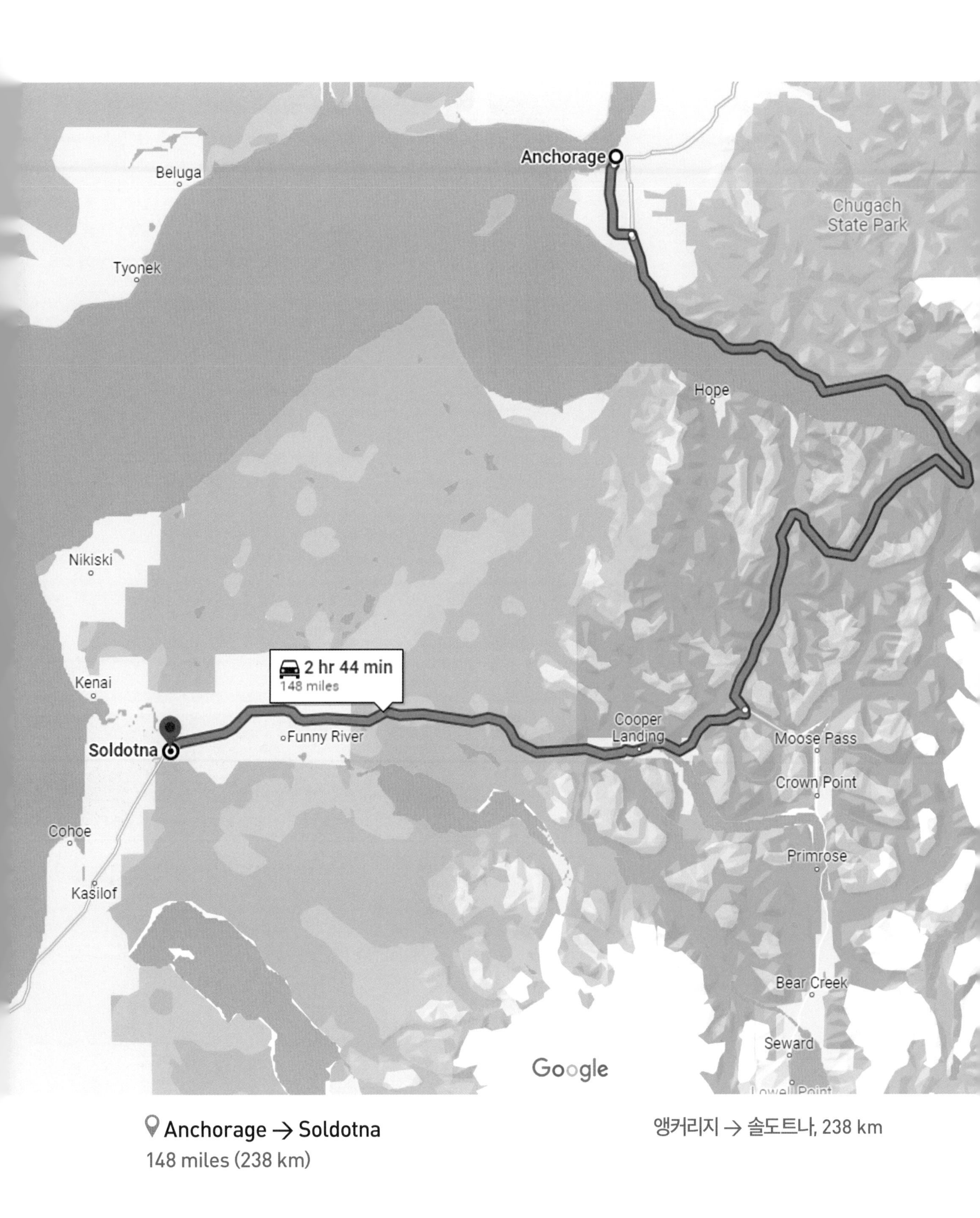

Anchorage → Soldotna
148 miles (238 km)

앵커리지 → 솔도트나, 238 km

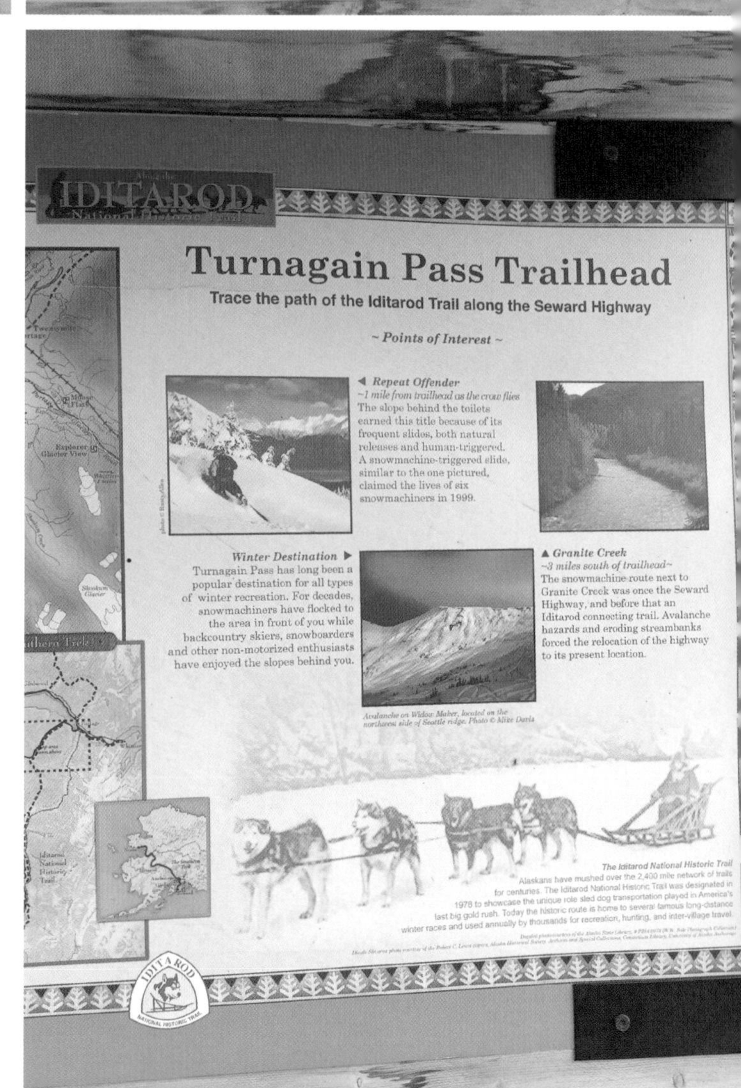

⊕⊘⊖

휴게소에 있는 광고물들

A sample of the signboards in the rest area

The picture I took in the rest area

휴게소에서 찍은 사진

The Alaska highway has
many straight roads
that appear endless.

알래스카 고속도로,
끝이 안 보이는 직선 길이 많다.

11th day

July 11th, Monday

11일째, 7월 11일 월요일

Finally we arrived at Ninilchik!

I met my old married friends after about 20 years. I was very pleased to see them again. While drinking coffee and tea, we talked about the distant past, allowing us to explore some deep thoughts. Furthermore, their son and daughters, who were elementary school students the last time I saw them, were now respectable young man and women.

Mattie, the youngest daughter had become a nurse and her husband was a firefighter. They now live with seven dogs and three horses on a forest estate of 10 acres (11,000 pyeong) on one side of a sea cliff. I envied them, because they seemed to be living the kind of life I had always wanted to live.

마침내 니닐칙에 도착!

약 20년 만에 옛날 친구 래리 부부를 만났다. 무척이나 반가워한다. 커피, 차를 마시며 오랜만에 지난 얘기를 하며 회포를 풀었다. 더구나 그때 당시 초등학생이던 아들과 딸들이 이제는 어엿한 청년과 숙녀들이 되어 있었다.

막내딸인 매티는 간호사가 되었고 그녀의 남편은 소방관이다. 그들은 10에이커(11,000평)의 나무 숲이 무성한 정원에 한쪽 면은 바다 절벽인 곳에서 개 7마리와 말 3마리와 함께 살고 있다. 내가 살고 싶었던 삶을 살고 있는 것 같아서 부러웠다.

Soldotna → Ninilchik
39 miles (63 km)

솔도트나 → 니닐칙, 63 km

Suzan and Larry's daughter
Mattie
and her husband Frank.

수잔과 래리의 딸 매티와 남편 프랭크

Carl and June meeting
the couple of Susan and Larry
for the first time in 20 years.

20여년 만에 만난 친구 래리와 수잔 부부

↑ ←

My friend's youngest daughter,
Mattie and her 7 dogs.
She's raising 3 horses too.

친구 막내 딸 매티와 애견 7마리.
말도 3마리 있다.

Some scenes
from the Ninilchik fish market.

니닐칙의 낚시배 대여와
잡은 고기를 정리해 주는 곳.

12th day

July 12th, Tuesday  12일째, 7월 12일 화요일

From today, I will be fishing for 3 days.

Today and the day after tomorrow, the plan is to go sea fishing for halibut, and tomorrow, I'll fish for sockeye salmon in the river.

I originally wanted to fish for king salmon (weighing an average of 20 pounds (10 kg)), but this year is a sabbatical year, so I can't catch them. Halibut are also restricted to just two per person.

오늘부터 3일간 낚시를 할 계획이다.
오늘과 모레는 핼리벗(광어) 낚시이고 내일은 강에서 삭아이 새먼(연어) 낚시다.
원래 킹새먼(보통 20 파운드(10kg))을 생각했는데 금년은 안식년이라서 못 잡는다고 한다. 핼리벗도 2마리로 제한되어 있다.

I just met a moose in town.

니널칙 동네에서 만난 무스.

Ⓘ

Here are Halibut we caught.

나와 가이드가 잡은 핼리벗

Ⓘ

My temporary cottage.
The accommodation
that my friend prepared for me to stay
at for the next 3 days.

친구가 마련해 줘서 3일간 머무른 숙소

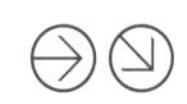

Scenes of my friend,
Larry and his brother, Frank
preparing halibut.

친구 래리와 그의 동생 프랭키가
핼리벗을 손질하는 장면

My friend's fishing boat.

친구의 낚시 보트

HH

I was enjoying halibut
and sockeye salmon sashimi.

회로 핼리벗과 삭아이 새먼을 즐겼다.

The other friends didn't eat
salmon and halibut sashimi.

다른 친구들은 생선을 회로 먹지 않는다.

A photo of Larry Corb and his family.
From left: Jeremy, Larry, Suzan, Mattie, Sarah.

래리 콥의 가족 사진.
(왼쪽부터) 제러미, 래리, 수잔, 매티, 세라.

13th day

July 13th, Wednesday

13일째, 7월 13일 수요일

Today we went river fishing.

I caught six sockeye salmon. Actually, I was limited to catching only six fish per day. Those restrictions don't apply to Alaska residents, only to outsiders.

My friend gutted the fish I had caught and prepared them to be eaten raw.

I enjoy eating sashimi (raw fish), but my friends here said, "No raw fish" for them, please.

오늘은 강 낚시다.

삭아이 새먼 6마리를 잡았다.

하루에 6마리까지만 잡을 수 있게 제한되어 있다. 알래스카 주민은 제한이 없지만 외지인은 적용시킨다. 내가 잡은 것을 회로 먹으라고 친구가 손질을 해줬다.

나는 회를 즐기는데 이 친구들은 날것을 못 먹는다고 했다.

A funny address sign

재미있는 집 주소 표지판

My cottage (cabin)

나와 준이 니닐칙에서 머물렀던 숙소

Alaska fishing permit

알래스카 낚시 허가증

Department of Fish and G
DIVISION OF ADMINISTRATIVE SER

THE STATE of ALASKA
GOVERNOR MIKE DUNLEAVY

Below is your 2022 Nonresident 1 Day Sport Fish License. If your license is not electronically signed must physically sign your license. You must carry the license in your possession whenever you are engaged in the licensed activity (sport fishing, hunting, etc.) and show it upon request to any person authorized to inspect it.

Please check your license carefully for accuracy. If there are any errors, email adfg.license@alaska call (907) 465-2376.

2022 NONRESIDENT 1 DAY SPORT FISH LICENSE #22868097
Effective 7/13/2022 12 PM to 7/14/2022 12 PM

KWANG Y AHN
29201 Camino Alba
Murrieta CA 92563-6604 USA

DOB: 3/23/1946
Driv Lic CA D5017973

Issued 07/13/2022

ALASKA FISH & WILDLIFE SAFEGUARD
(800) 478-3377
Callers remain anonymous and are eligible for a reward

Questions: adfg.license@alaska.gov or (907)465-2376

Fee may include surcharges per A.S. 16.05.340

I have read and understand the definition of Alaska "resident" (AS 16 AS 16.05.940). My license has not been suspended or revoked in an information is subject to public disclosure. Making false statements to criminal penalties (AS 11.56.210 and AS 16.05.420). **I certify all is true and correct.**

Signature of Licensee X

The following species have annual limits in specific areas. Check the current sport fishing regulations for current annual limit provisions.

1. **King Salmon** - Fresh waters: Kenai Peninsula, Susitna-West Cook Inlet, Bristol Bay, Unalakleet River, Aniak River, Upper Copper River, Kodiak Island, Alaska Peninsula, Aleutian Islands, Southeast Alaska (nonresidents only) and Yakutat (nonresidents only); Salt waters: Cook Inlet, Southeast Alaska (nonresidents only), and Yakutat (nonresidents only).

2. **Steelhead/Rainbow Trout** - Kenai Peninsula, Susitna-West Cook Inlet, Southeast Alaska, Yakutat, Alaska Peninsula/Aleutian Islands, Kodiak Island, Prince William Sound, and Lower Kuskokwim River drainages.
3. **Lingcod** - Southeast Alaska (nonresidents only).
4. **Sablefish** (Black Cod) - Southeast Alaska (nonresidents only).
5. **Yelloweye Rockfish** - Southeast Alaska (nonresidents only).
6. **Sharks** (except spiny dogfish) - Statewide.
7. **Other**

Check current regulations for harvest limits. Immediately upon harvesting a fish for which an annual limit exists, you are required to record details (in ink) below.

Water	Species	Date

Fishing maniacs

낚시를 즐기는 강태공들

Seagulls are waiting for the salmon's gutting.

연어 내장 제거를 기다리는 갈매기들

Fishingholics

못말리는 낚시광들

The sockeye salmon I caught.
(also known as red salmon.)

내가 잡은 삭아이 새먼
(홍연어라고도 함.)

14th day

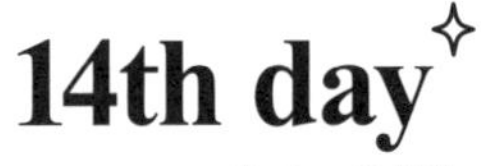

July 14th, Thursday

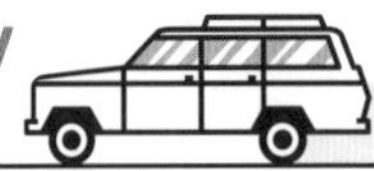

14일째, 7월 14일 목요일

After two days of fishing, my shoulders and arms are achy, and I experienced a throbbing pain. I decided to cancel today's fishing schedule and instead lie down and rest, but June was eager to go out.

So, I walked around the neighborhood with June.

We arrived in the city of Homer, 40 miles south of Ninilchik. Seeing the "LAND'S END" sign here is impressive.

We found a Korean-run Japanese restaurant. The owner made me a spicy mixed-seafood stew that really suited my taste. It was so delicious that I ate two bowls of rice with it.

Tomorrow, I have to go back, following the same route from which we came.

2일 동안 무리하게 낚시를 했더니 어깨와 팔이 아프고 욱신거린다.
오늘 예약한 낚시 일정을 취소하고 느긋하게 누워 좀 쉬려고 하는데 옆에서 준이 나가자고 보챈다.
준과 동네 한 바퀴를 돌았다.
니닐칙에서 남쪽으로 40마일 떨어져 있는 호머라는 도시에 도착했다.
땅끝이라는 표지석이 인상적이다. 이곳에서 한국인이 운영하는 일식집을 발견. 주인장이 내 입맛에 맞게 얼큰한 해물잡탕을 만들어주었다. 꿀맛이어서 밥 두 그릇을 뚝딱 먹었다. 내일부터는 왔던 길을 되돌아 가야 한다.

↑
Spicy jjamppong. Rice, kimchi, etc. are not on the menu, but they made it specifically for me. It was so delicious.

얼큰한 짬뽕. 밥, 김치 등 메뉴에도 없는데 해준다. 꿀맛이었다.

LAND'S
END

ⓘ

The Homer Ferry Terminal

호머의 페리 정착장

The Land's End signboard.

땅끝 표지석

Ⓣ

My friend, Larry Corb and his brother, Franky.

친구 래리 콥과 그의 동생 프랭키

Frank's handmade machine for smoking salmon.

프랭크가 손수 만든 연어를 훈제하는 기계

15th day

July 15th, Friday

15일째, 7월 15일 금요일

Finally, we're beginning our returning today.
At 9 a.m., I left my beloved cabin, in which I'd spent the past 4 nights and 5 days, making sure to fill up on gas at the only gas station in Ninilchik. We had a late breakfast consisting of yesterday's leftovers, and then hit the road. Once I get to a bit bigger neighborhood along our route, I'm going to stop and get a check up on our vehicle.
It has been raining since yesterday. I was in bad physical condition because I got rained on from time to time during our drive, so even though we could have continued on, I decided to stop over in the city of Soldotna and go to a motel a little early to rest. We were actually very tired. June also lost a lot of weight.
I guess he's having a hard time, too …

드디어 오늘 돌아간다.
4박 5일 머물던 정든 오두막을 오전 9시에 떠나서 니닐칙에 하나뿐인 주유소에서 주유하고 어제 남겨온 남은 음식으로 늦은 아침을 먹고는 먼 길을 나섰다. 가는 도중에 좀 더 큰 동네를 만나면 차를 점검할 예정이다.
어제부터 비가 온다. 오는 중간에 간간이 비를 맞았더니 컨디션이 좋지 않아서, 마음 같아선 더 갈 수도 있겠는데 솔도트나에서 숙박하기로 마음먹고 좀 이른 시간에 모텔에 들어가 쉬기로 했다. 우리는 사실 많이 지쳐 있었다. 준은 체중도 많이 줄었다.
이 녀석도 힘든가 보다….

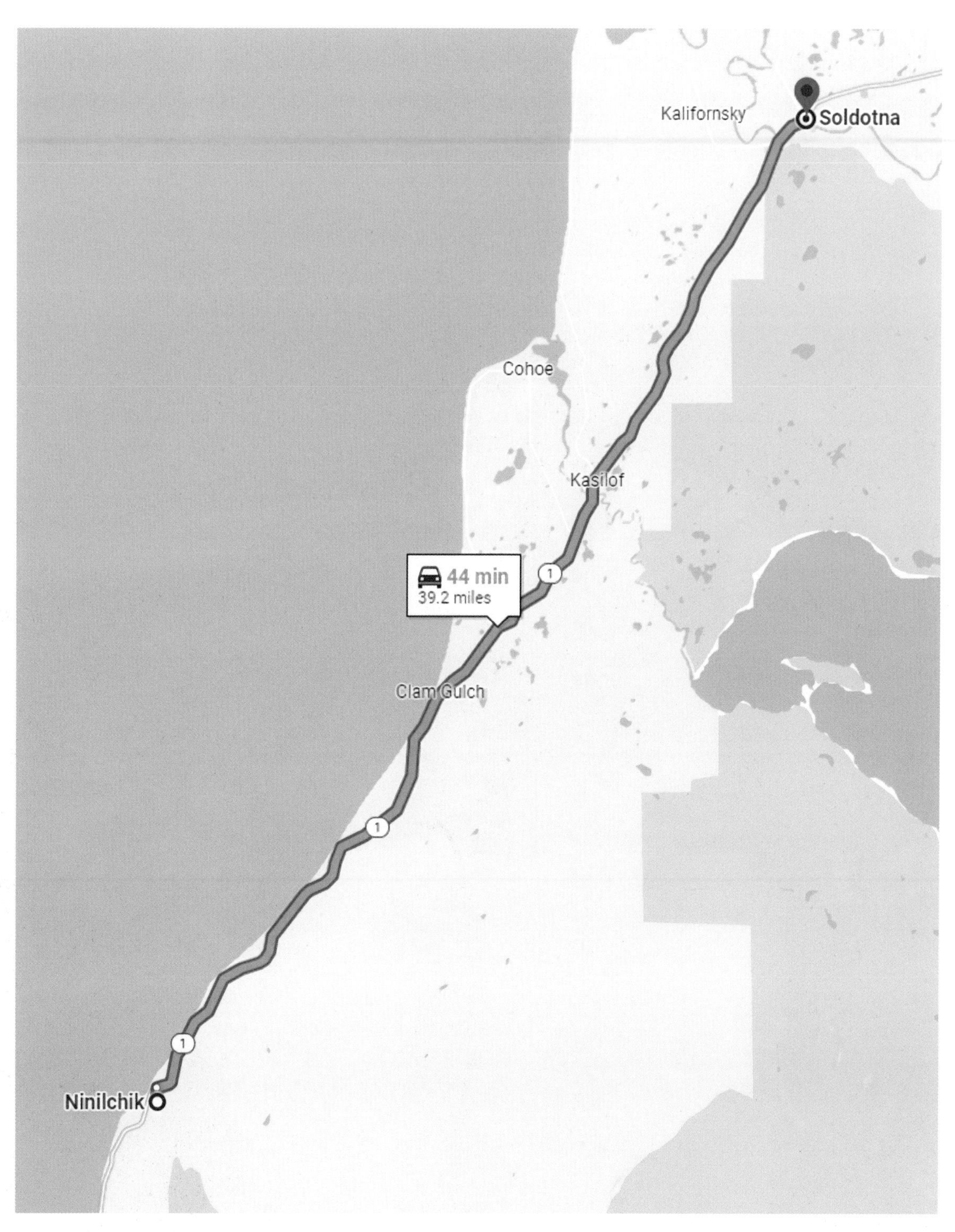

Ninilchik → Soldotna
39 miles (63 km)

니닐칙 → 솔도트나, 63 km

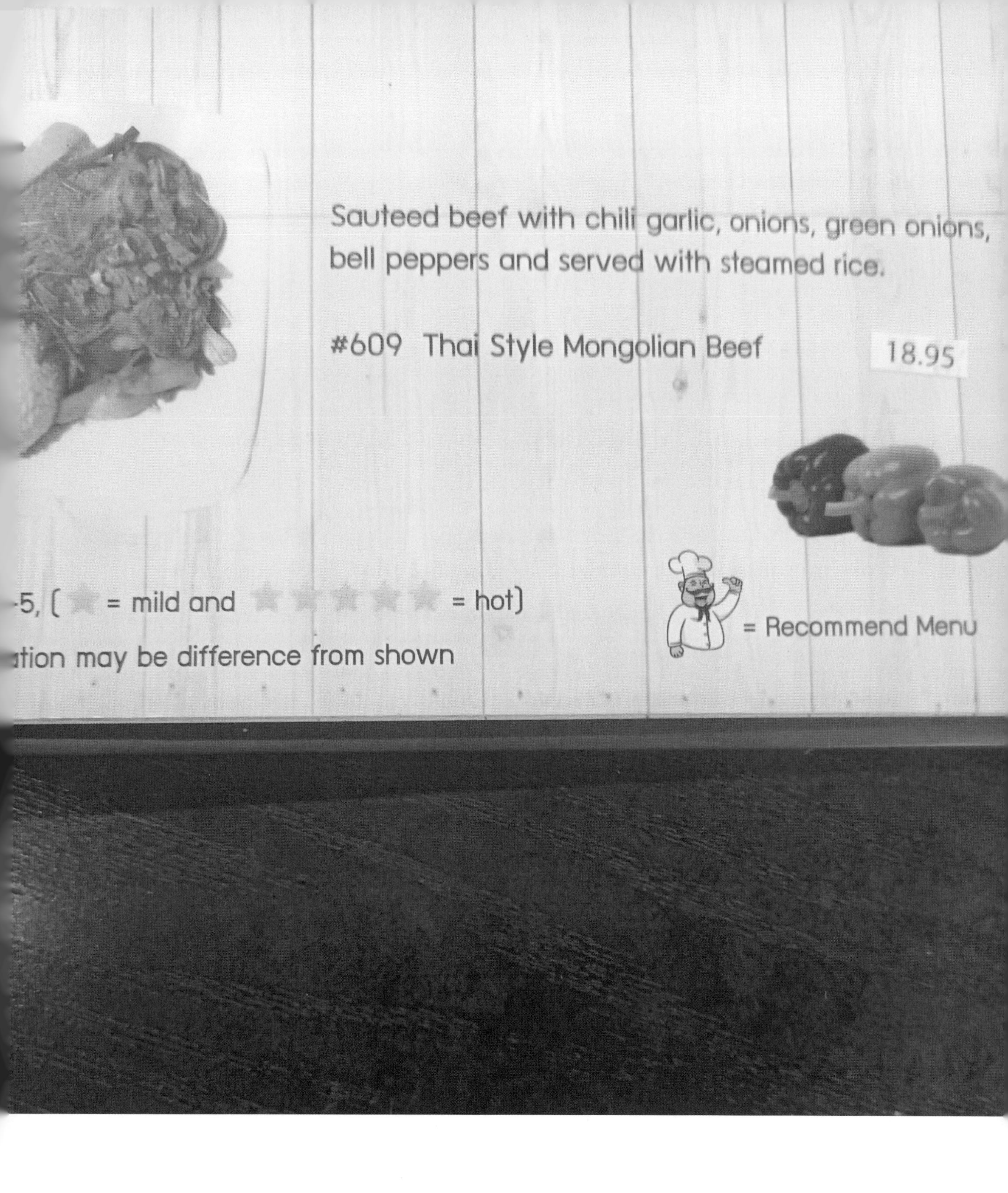

Sauteed beef with chili garlic, onions, green onions,
bell peppers and served with steamed rice.
#609 Thai Style Mongolian Beef
18.95
-5, (= mild and = hot)
ation may be difference from shown
= Recommend Menu

The Thai style Mongolian beef we ate at a Thai restaurant
in Soldotna. What can I say, spicy and salty things just suit my taste.

솔도트나에 있는 타이 레스토랑에서 먹은 타이 스타일 몽골리안 비프.
그래도 매운 맛이 내 입맛에 맞는다.

16th day

July 16th, Saturday

16일째, 7월 16일 토요일

It is the second day of our return.
About an hour ago, I was dreaming in a deep sleep, but then June hit my stomach with his two paws to wake me up. I guess he was in a hurry to go to the bathroom.
I woke up with a sore throat and phlegm. I'm worried. I still have to drive for 10 days … If I go to Anchorage, I should buy some medicine at the first chance I get.
In the evening, I met up for my scheduled appointment with Lillianne, and we ordered galbi and soybean paste stew at a Korean restaurant. We ate with great pleasure. June did too, as he got the leftover ribs.

되돌아가기 2일째.
1시간 전 한참 꿈을 꾸며 자고 있었는데 준이 두발로 내 배를 덮쳐 눌러 잠이 깼다. 화장실이 급한 모양이었다.
일어나니 목이 아프고 가래가 생겼다. 걱정이 된다. 아직 10일은 운전해서 가야 하는데…. 앵커리지에 가면 약부터 사서 먹어야겠다.
저녁식사를 약속한 릴리안을 만나 한국식당에서 갈비랑 된장찌개를 시켜 맛있게 먹고는 남겨온 갈비를 준에게도 나눠줬다.

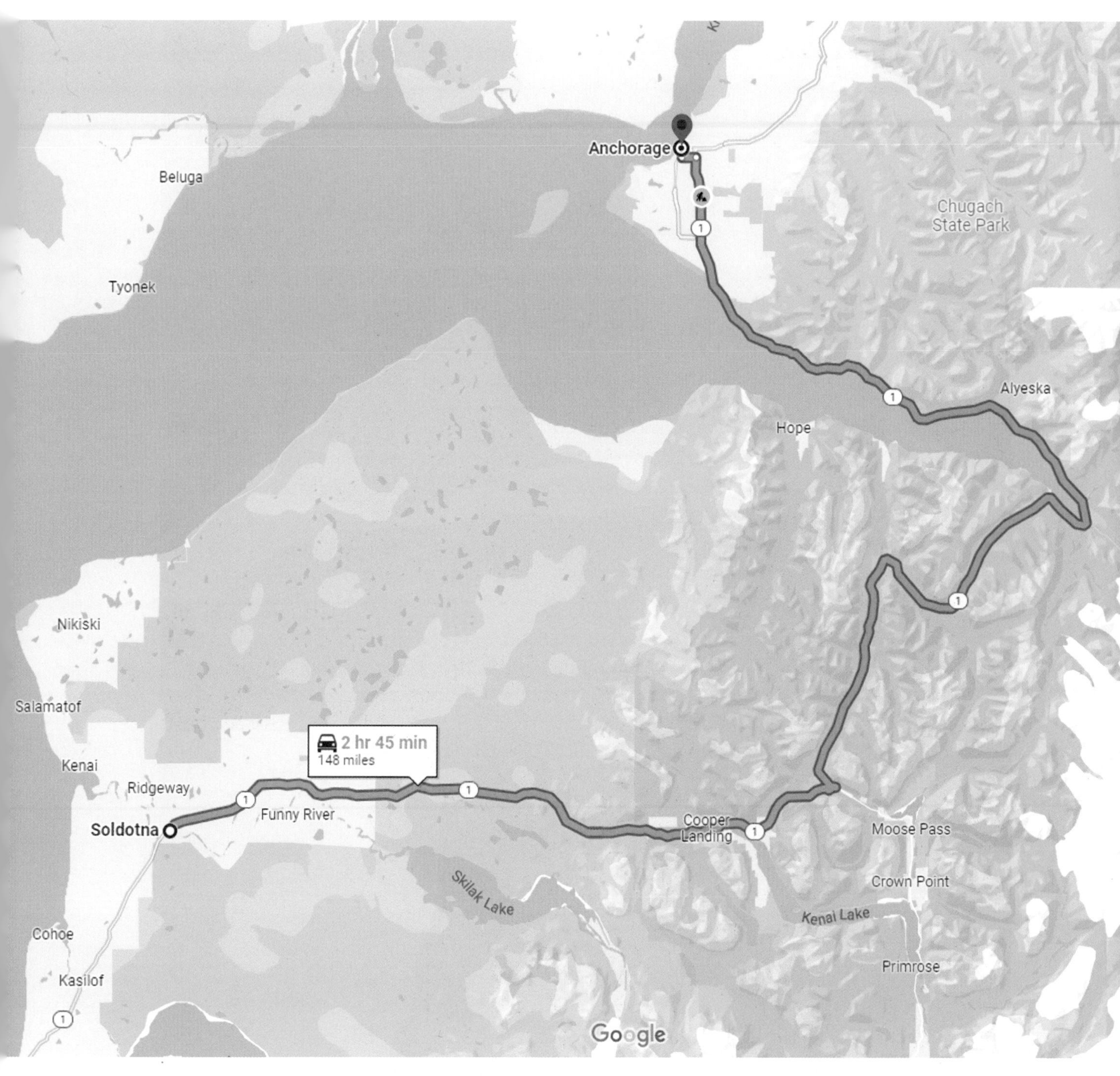

Soldotna → Anchorage
148 miles (238 km)

솔도트나 → 앵커리지, 238 km

⊕

Exhausted June

기진맥진한 준

⊖

Lillianne's Toyota passenger car.
She has her summer house attached onto its rooftop.

다시 만난 릴리안의 여름별장

VALHALLA
Center
ALASKA
JLX 905

17th day

17일째, 7월 17일 일요일

The third day of our return.

I'm planning to go to Tok today.

Today is the third day of rain here in Alaska.

I bought medicine to treat my sore throat and phlegm at a pharmacy.

I saw the sign for "Subway" from the highway, so I went in and ate a late lunch there, before continuing our drive. Subway sandwiches are more palatable to me than hamburgers.

We're now 100 miles from the Canadian border crossing point. We are staying in a small town called Tok next to a river in the mountains.

되돌아가기 3일째.

오늘은 톡까지 가는 것으로 계획.

이곳 알래스카에는 3일째 비가 내린다.

약국에서 목 아픔과 가래 삭히는 약을 사서 복용했다.

고속도로에서 '서브웨이' 사인이 보여 찾아 들어가 그곳에서 늦은 점심을 먹고는 계속 달렸다. 햄버거보다는 서브웨이 샌드위치가 내 입맛에 맞는다.

캐나다 국경 100여 마일 지점. 산속 강 옆에 있는 톡이라는 작은 동네에서 숙박.

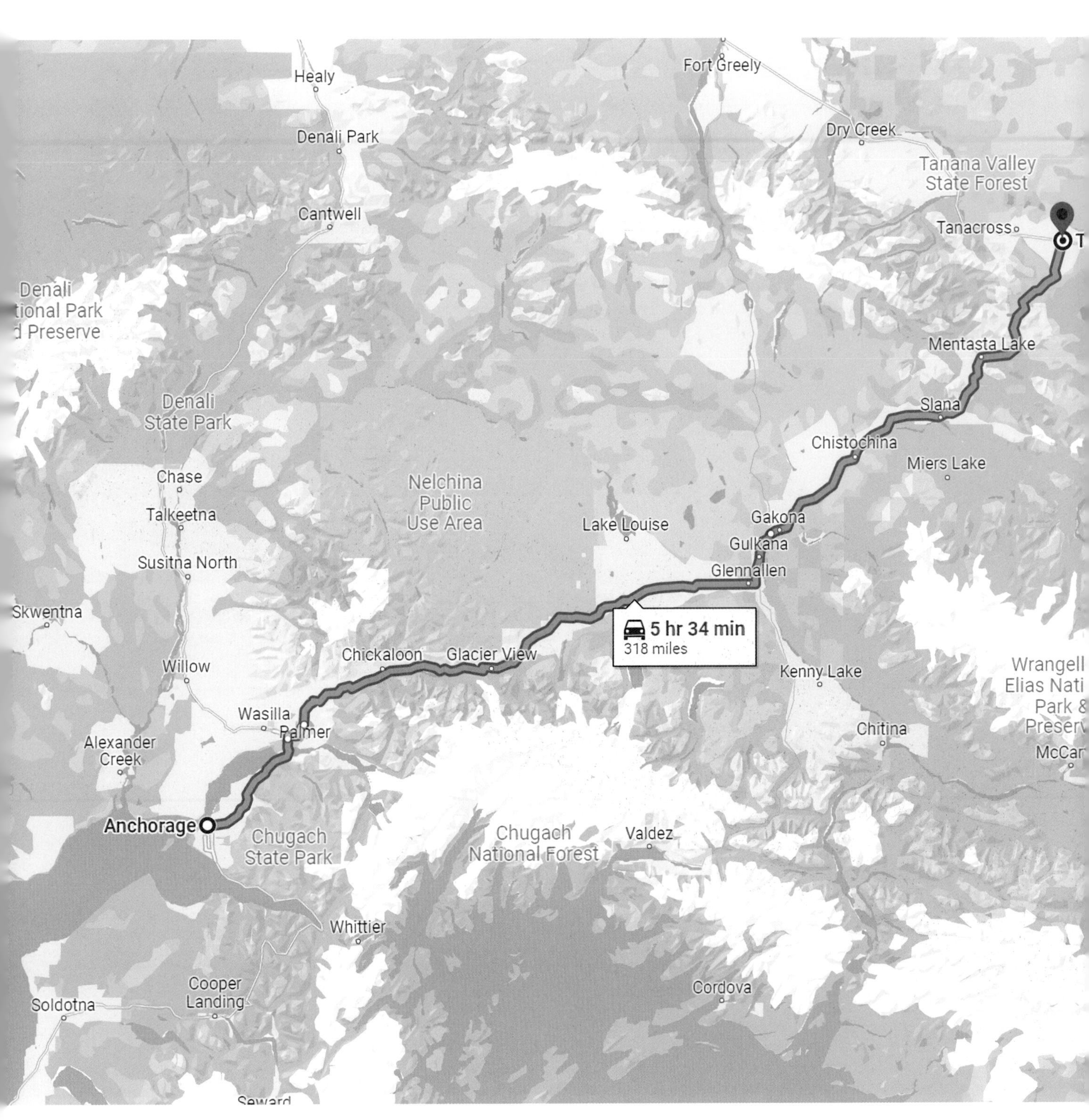

Anchorage → Tok
318 miles 509 km

앵커리지 → 톡, 509 km

Subway beside the highway in a small town.

고속도로에서 만난 반가운 서브웨이

This dinner special included the same kind of sockeye salmon that I caught, so I ordered it.

내가 잡았었던 삭아이 새먼이
디너 스페셜 메뉴에 있어서 시켰다.

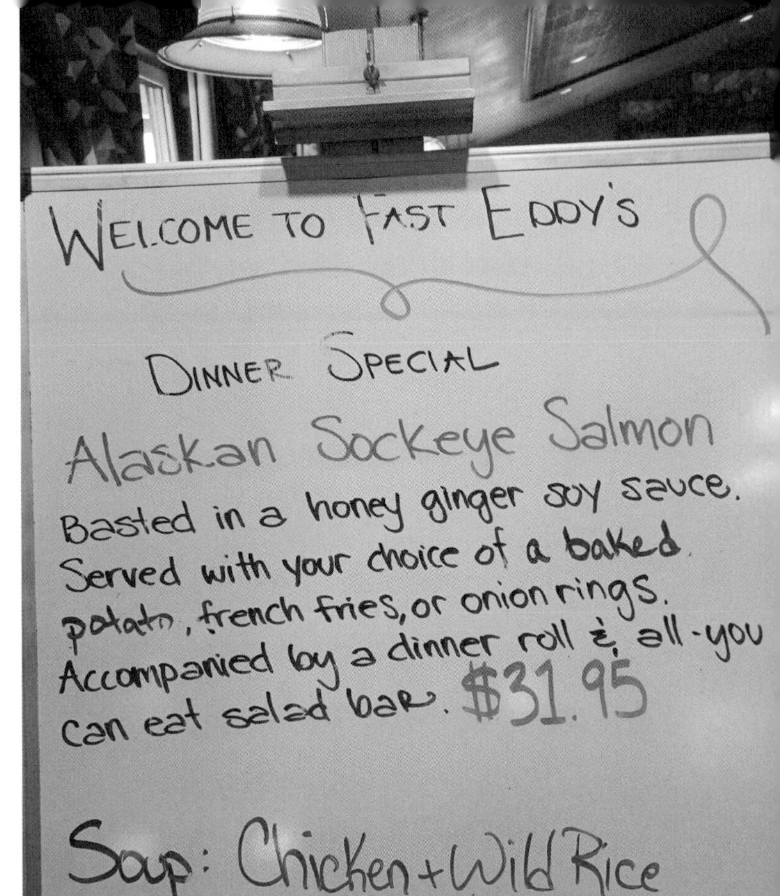

WELCOME TO FAST EDDY'S
DINNER SPECIAL
Alaskan Sockeye Salmon
Basted in a honey ginger soy sauce.
Served with your choice of a baked potato, french fries, or onion rings.
Accompanied by a dinner roll & all-you can eat salad bar. $31.95
Soup: Chicken + Wild Rice

However, the taste was nothing special.
I passed it on to June.

그런데 맛은 별로였다.
준에게 갖다 주었다.

18th day

July 18th Monday

18일째, 7월 18일 월요일

The 4th day of our return.

Guess who woke me up today from a sound sleep? Yep, it was June again. It's early in the morning, right around dawn.

We are crossing the border today. From the Yukon Province, it's a 3-day drive through the Rockies into British Columbia. After passing through the Rockies, I will arrive in Vancouver on the fourth day. The only thought on my mind right now is, "If you get to Vancouver, you'll be able to eat Korean food again". Yes, I miss kimchi and rice. I'm a 77-years young Korean.

We're staying at the Gold Rush Inn tonight.

되돌아가기 4일째.

곤하게 자고 있는 나를 깨우는 것은 역시 준이다. 이른 아침, 아직 새벽이다.

오늘 국경을 넘는다. 유콘 자치주 록키를 3일 달리면 브리티쉬 콜럼비아 주에 들어간다.

그곳을 지나서 4일 후면 밴쿠버에 도착할 것이다. 밴쿠버에 가면 한국 음식을 먹을 수 있으리라. 오로지 그 생각만 하고 있다.

김치와 밥이 먹고 싶다. 나는 77살 젊은 한국인이다.

골드 러쉬 인에서 숙박.

Tok → Whitehorse
386 miles (622 km)

톡 → 화이트 호스, 622 km

NO
PARKING
ANY
TIME

June, enjoying the view.

경치를 즐기는 준

You can see many mountains covered in snow at the top.

멀리 눈 덮힌 아름다운 산이 보인다

Wildlife Refuge?
I left the car windows down while going to the bathroom here, and was surprised to find several dozens of mosquitoes in the car when I returned. It took me half an hour to eliminate them all.

와일드라이프 래퓨지? 창문을 열어놓고 화장실에 다녀왔더니 차 안에 모기들이 수십 마리 들어와 있어서 깜짝 놀랐다. 퇴치하는데 30분이나 소요됐다.

※편집자 주: 야생동물 보호구역을 뜻합니다.

Ⓘ

Since there were no Korean or Japanese restaurants here,
I visited a Chinese restaurant.
I found kimchi on the menu and ordered it.
It costs 8 dollars and 50 cents.
Unlike shown on the picture, the amount was small.

한국식당이나 일본식당도 없어 찾은 중국식당. 메뉴에서 김치를 찾아 시켰다. 8불 50센트를 받는다.
사진과 다르게 양이 적었다.

Ⓘ

While returning, we stopped in front of another lakeside rest area with a lightweight plane.

돌아가면서 다시 들른 경비행기가 있던 휴게소 호수 앞에서.

19th day

19일째, 7월 19일 화요일

It is the 5th day of our return.

I woke up at 6 a.m, and June, who always wakes me up around this time, was still lying in bed. Is this little guy sick, too? I'm speechless … and worried.

From today, we're going to be driving non-stop for three days through the Rocky Mountains range to reach Vancouver. Please God, bless me & June!

There are many straight roads heading straight into the horizon, and no one says anything even if you exceed 100 miles per hour (160 ㎞/h). There are not many cars on the road and no traffic police. Sometimes you pass by the occasional motorcycle or bicycle. You can even see people crossing on foot.

되돌아가기 5일째. 눈을 뜨니 새벽 6시, 평소 이 시간이면 항상 나를 깨우던 준이 아직도 침대에 누워있다. 이 녀석도 어디 아픈가? 말도 못하고… 걱정이 된다.

오늘부터 3일간 록키 산맥을 열심히 달려야 밴쿠버까지 갈 수 있다. 참으로 광활한 엄청 큰 나라. 신이여, 준과 나에게 은총을!

지평선이 눈앞에 보이는 직선 길이 많고 100마일(160km)로 달려도 누가 뭐라 하는 사람이 없다. 자동차도 별로 다니지 않고 교통경찰도 없다. 간간이 오토바이와 자전거로 통과하거나 또는 도보 횡단하는 사람들이 보이기도 한다.

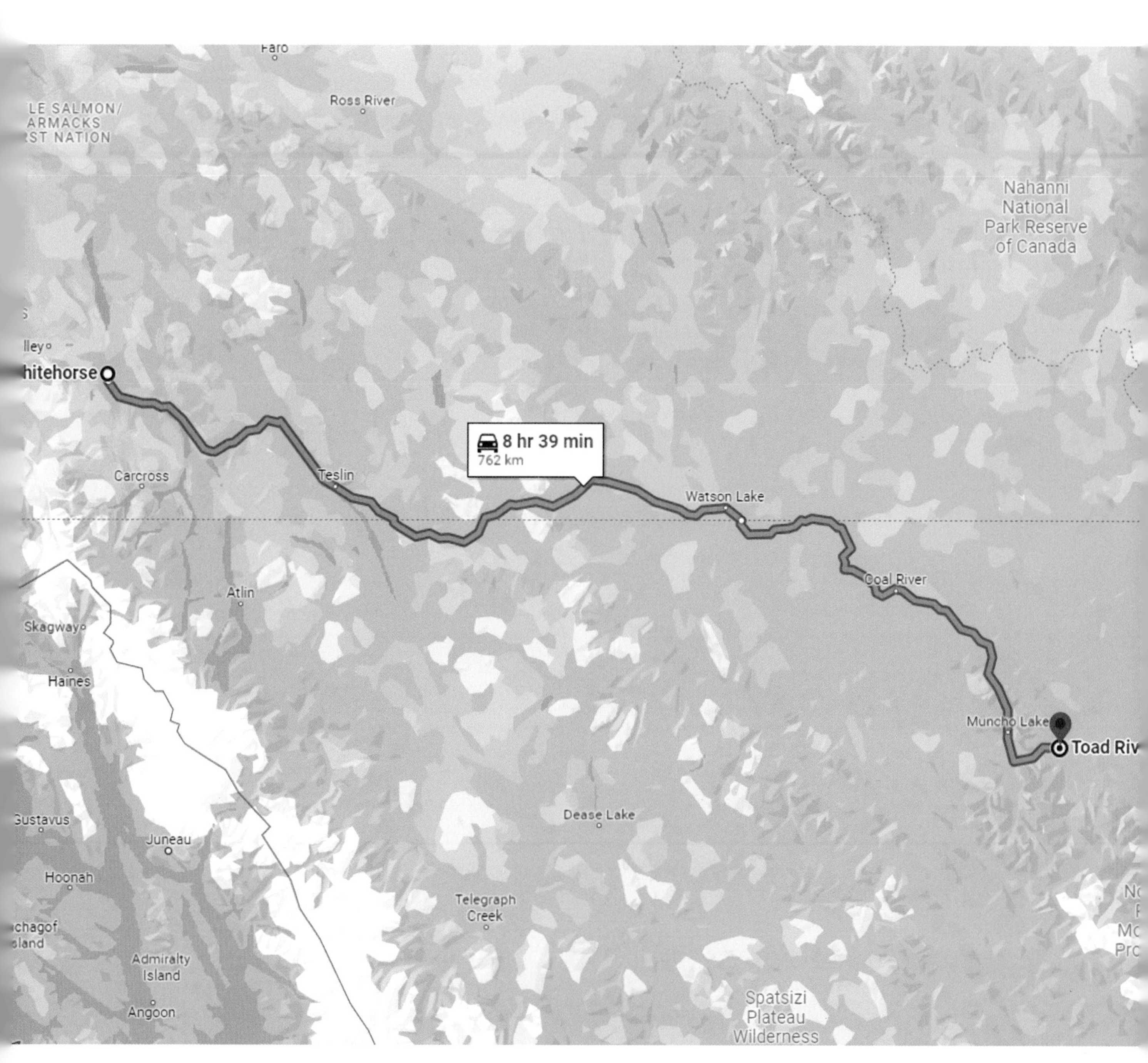

Whitehorse → Toad River
473 miles (762 km)

화이트 호스 → 토드 리버, 762 km

140
160
OFF
102
4:33

⊖⊘

At a rest area

한적한 휴게소에서

⊖

Along the Alaska Highway,
you can see
a beautiful flower path
stretching along both sides
of the road. If you look up
and down the road,
there's not another car in sight.
I've driven at speeds of up to
140 miles per hour (224 *km*).

알래스카 고속도로 양 옆으로 계속 보이는
아름다운 꽃길. 앞 뒤로 차가 보이지 않는다.
140마일(224*km*)까지 밟아 봤다.

⊖

The first rest area
after leaving White Horse.

화이트 호스 후 첫 번째 휴게소

⊖⊕

The toilet at a rest area
along the Alaska Hwy.
There wasn't any smell at all.

알래스카 고속도로 휴게소 화장실.
보기와 달리 냄새가 없다.

20th day

July 20th, Wednesday

20일째, 7월 20일 수요일

It is the 6th day of our return.
We stayed at the Toad River Lodge, located deep in the mountains of the Rockies. Along the way, we saw three baby bears, several buffaloes, sheep, and many wild animals whose names I didn't know.
The nightfall came quickly, and I had to rush through the course without taking many pictures, because I was afraid I'd arrive at the accommodation too late.
We passed through the Canadian Yukon territory to enter the state of British Columbia.
I have to get a sound sleep deep in the mountains tonight.
Thank God!

되돌아가기 6일째.
록키 산맥의 깊은 산속 강가에 위치한 토드 리버 롯지.
오는 길에 아기 곰 3마리와 여러 마리의 버팔로와 양들 그리고 이름 모르는 야생동물들을 많이 보았다.
날은 어두워지고 숙소에 늦을까봐 사진도 못 찍고 빨리 달려야만 했다.
캐나다 자치주 유콘에서 브리티쉬 콜럼비아 주로 넘어왔다.
오늘밤도 깊고 깊은 산속에서 자야 한다.
신이여, 감사합니다!

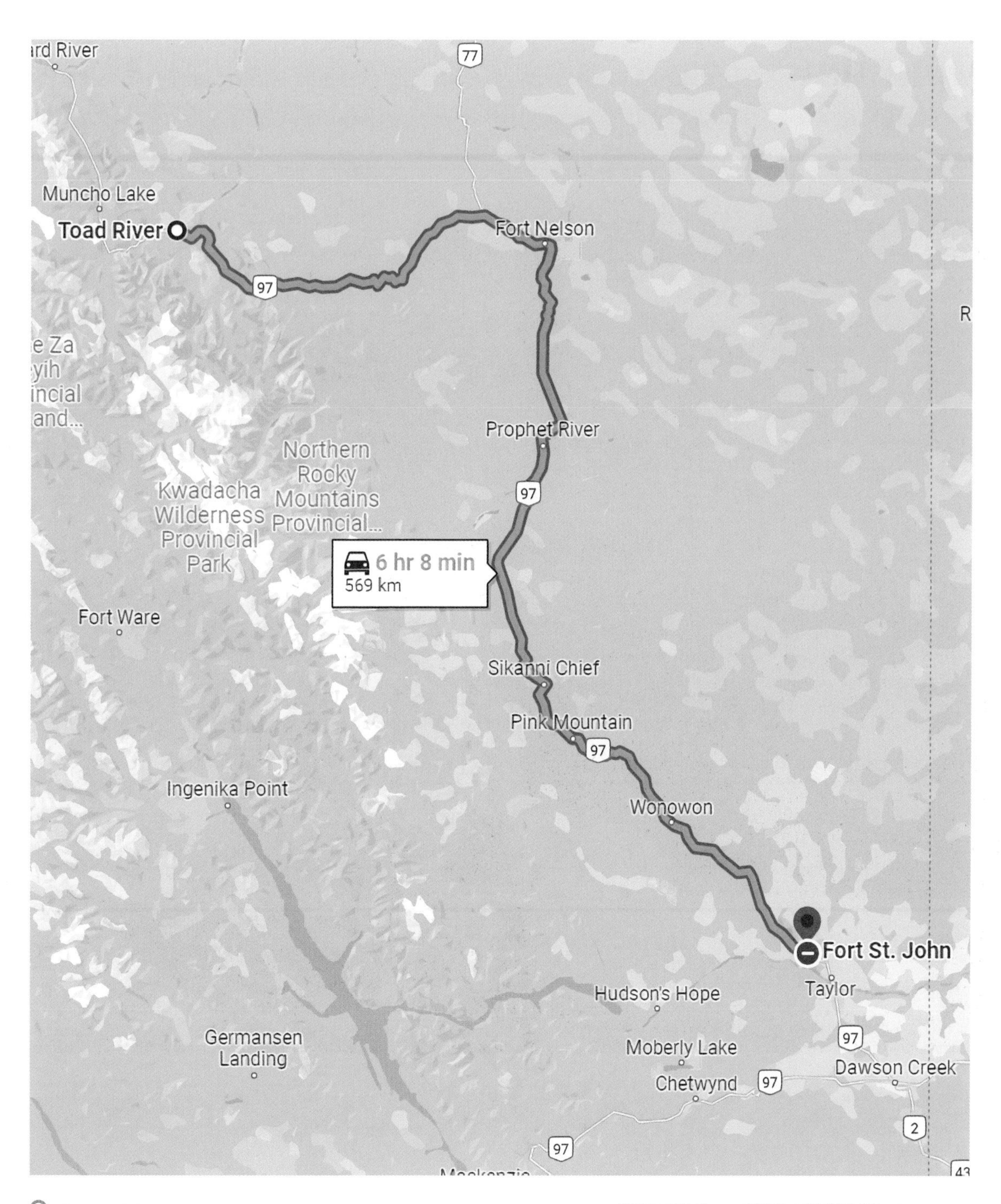

Toad River → Fort St. John
353 miles (569 km)

토드 리버 → 포트 세인트 존, 569 km

A photo from my lodging

숙소에서 찍은 사진

June is enjoying the view. It was just beautiful!

무슨 생각을 하니, 준! 아름다움을 아는거니?

21st day

July 21th, Thursday

21일째, 7월 21일 목요일

It has been a week into our return journey.

June woke me up at 6:30.

Thanks, June.

Today, we need to drive as much as we did yesterday, around 400 miles, to arrive at Williams Lake. I've reserved the accommodation in advance.

This fall semester, I've decided I definitely need to register for some kind of smart phone and computer literacy classes. There are so many inconveniences I experience because I am not good at these two things.

I would like to thank all the people who have been praying for the success of this trip.

Thank Jesus!

되돌아가기 일주일째.

준이 6시 30분에 깨워줘서 일어났다. 땡큐, 준.

오늘도 어제만큼 산속 길을 400여 마일 달려야지만 윌리엄스 레이크에 도착할 수 있다. 미리 숙소를 예약해 두었다.

가을 학기에는 꼭 스마트폰과 컴퓨터 클래스를 등록할 예정이다. 이 두 가지를 잘 못하니 불편한 것이 너무 많다.

이번 여행의 성공을 기도해 주시는 많은 분들에게 감사를 보낸다.

예수님, 감사합니다!

Fort St. John → Williams Lake
420 miles (676 km)

포트 세인트 존 → 윌리엄스 레이크, 676 km

DIESEL

June in the back seat of my car.

내 차 뒷좌석을 차지한 준

A quiet rest area without cars.

잠깐 쉬어 갈까?

A signboard at a rest area.

휴게소에 있는 광고판들

A tired-looking June.

나보다 더 피곤해보이는 준

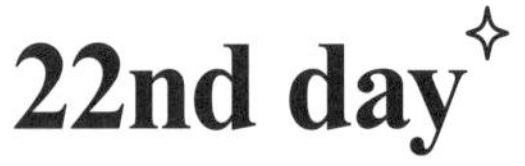

22nd day

July 22nd, Friday 22일째, 7월 22일 금요일

Today is the 8th day of our return, and 22 days overall since I began this battle with my mental resolution and stamina.

Yesterday I set the alarm before going to sleep. I awoke at 5 a.m.!

We need to drive more than 300 miles today if we are to reach Vancouver as planned along the Trans-Canada highway.

We've been driving through mountains for 4 days now, but today was the highlight. For about three hours, there was snow at the top, me and June in the middle, and many people rafting in the river at the bottom of the valley, while alongside, a train with no front and no end was moving beside the river.

The Canadian trains seem to be longer than the American ones.

God bless me & June, please!

나의 정신력과 체력과의 싸움 22일째, 되돌아가기 8일째.

어제는 알람 설정을 해 놓고 잤다.

5시 기상!

오늘 3백 마일 이상을 달려야 밴쿠버에 도착한다.

트랜스 캐나다 하이웨이.

4일간 산 속을 달렸지만 오늘 오후의 경치가 백미였다.

대략 3시간 동안 정상에는 흰 눈이, 중턱에는 내가, 계곡 밑으로 강에는 래프팅하는 인파들이, 강 옆으로 앞과 끝이 안 보이는 기차가 지나가고 있었다.

캐나다 기차가 미국 기차보다 더 긴 것 같다.

신이여, 저와 준에게 안전을!

Williams Lake → Vancouver
341 miles (549 km)

윌리엄스 레이크 → 밴쿠버, 549 km

A signboard explaining about a famous town during the gold rush.

황금 채취로 유명했다는 마을 안내

Ⓘ↑ ↗ →

A resort that combines rafting and accommodation.

래프팅과 숙박을 겸하는 리조트

23rd day

July 23th, Saturday

23일째, 7월 23일 토요일

It is the 9th day of our return.

I entered the motel last night and lay down on my bed, soon falling asleep without even showering. When I woke up, it was 3 a.m. and I'd been sleeping so deeply for 5-6 hours that you'd think I were dead.

Perhaps that's why I felt light-hearted today …

Today we crossed the Canadian-U.S. border. There are about 1,200 miles left before we reach L.A.

Tonight's accommodation is in Albany, Oregon.

되돌아가기 9일째.

어제 모텔에 들어와 잠깐 누워 있다가 샤워한다는 것이 그냥 잠이 들어버렸다.

깨어보니 새벽 3시, 5~6시간을 죽은 듯이 깊은 잠에 빠졌던 것 같다.

그래서인지 몸은 가뿐했다.

오늘 국경(캐나다 — 미국)을 넘었다. 엘에이까지 약 1,200마일 남았다.

오리건 주 올버니에서 숙박.

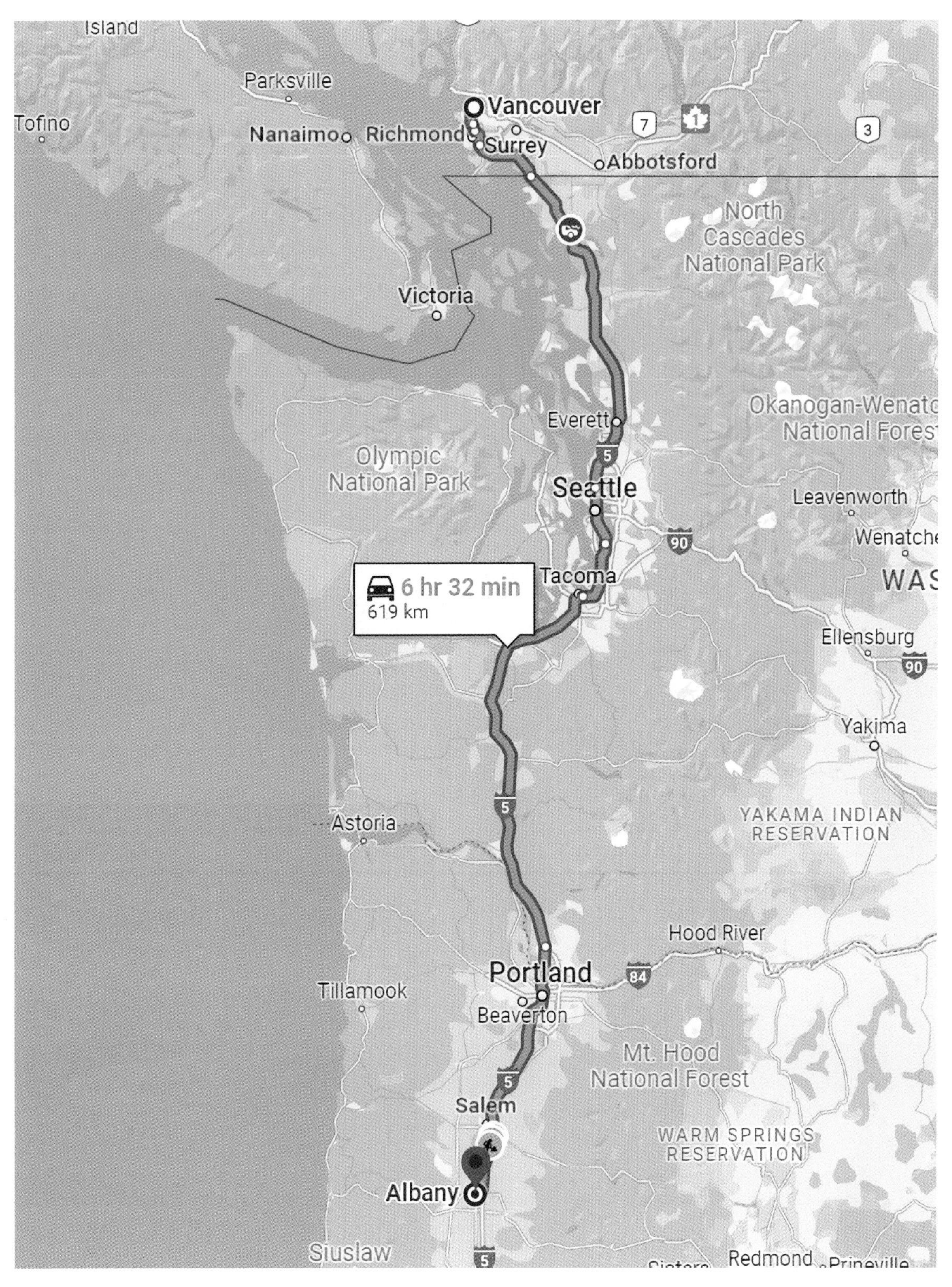

Vancouver → Albany
384 miles (619 km)

밴쿠버 → 올버니, 619 km

My smart boy June.

스마트한 나의 준

A Korean restaurant that was closed because of Covid19.

코비드19 때문에 문을 닫은 한국 식당

KIMCHI Restaurant
KOREAN & JAPANESE
Licensed

24th day

July 24h, Sunday

24일째, 7월 24일 일요일

It is the 10th day of our homecoming war.

Yesterday was the longest day so far … 2 hours waiting to cross the border from Vancouver … another 2 hours spent on the highway from Seattle, stuck there because of a roadside accident …

We visited Portland, and went to the "Portland Japanese Garden", which I thought was a good Japanese restaurant where I had planned to have dinner. To my surpride, I arrived to discover it was a real Japanese garden, created by the Japanese government. That was a bit of a letdown …

So I ended up skipping both lunch and dinner.

Tonight, checking in at the town of Williams, near San Francisco.

※In Korea, they sometimes call a restaurant a garden.

귀가 전쟁 10일째.

어제가 제일 긴 하루였다. 밴쿠버 국경 넘는데 2시간 소요,

시애틀 고속도로 사고 정체로 도로에서 2시간 소요.

포틀랜드에서는 저녁을 먹으려고 괜찮은 일식집으로 알고 찾아간 '포틀랜드 재패니즈 가든'이 도착해서 보니 일본정부가 만들어 놓은 진짜 일본식 정원이었다니!! 기대했는데 허탈했다.

그래서 점심도 저녁도 못 먹었다. 오늘은 샌프란시스코 근처 윌리엄스 체크인…

※한국에서는 식당을 가든이라고도 함.

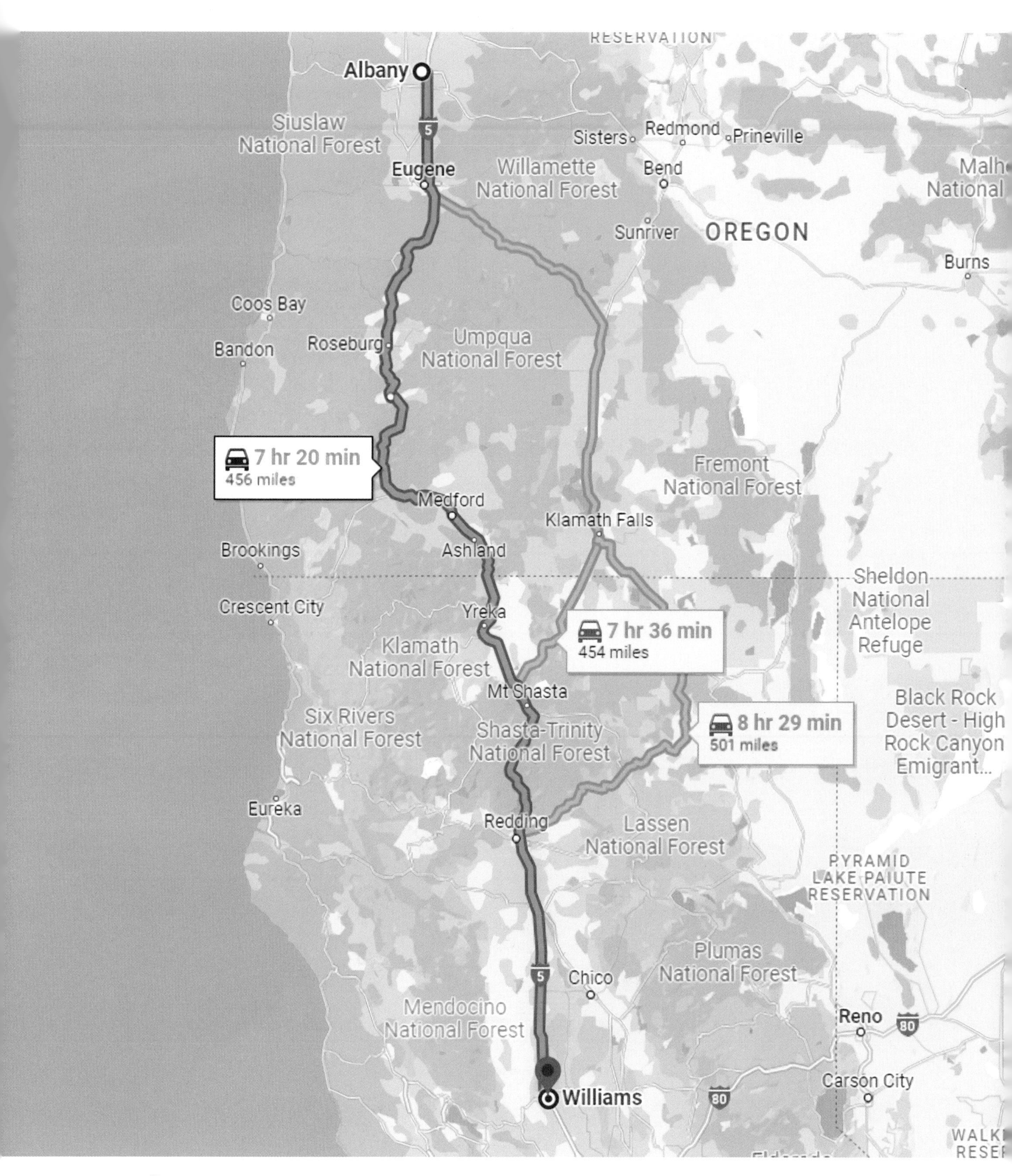

Albany → Willams
456 miles (733 km)

올버니 → 윌리엄스, 733 km

The Portland Japanese Garden.

포틀랜드에 있는 일본 정원.

The morning sunrise, reflected from the motel window.

숙소 창문을 통해 본 아침 해뜨는 장면

25th day

July 25th Monday

25일째, 7월 25일 월요일

It is the 11th day of our return.

My alarm clock, June! He smacked me with his two feet at 6 a.m. this morning. If he decides to do that near my face one day, I'm really going to get hurt … that little rascal …

Today, I will visit my friend Jo who lives in Ojai.

As I mentioned, I visited Jo's house and it was in a deep mountain village. Together with Jo and her dog, Rocky, the four of us had a nice dinner and tea, as well as some good conversation. Jo worked for Alaska Airlines for a long time, and is now volunteering to aid early-onset dementia patients.

되돌아가기 11일째.

준은 나의 알람시계! 새벽 6시, 자고 있던 나를 두 발로 덮친다. 얼굴 쪽으로 덮치면 내가 다칠 텐데 기특한 녀석…. 오늘 예정은 오하이에 사는 친구 조를 방문할 예정이다.

알려준 대로 집을 찾아가니 첩첩산중에 있다. 조와 그녀의 반려견 록키와 함께 넷이서 저녁을 맛있게 먹고 차를 마시며 많은 얘기도 나누었다.

조는 알래스카 에어라인에서 오랫동안 근무를 했고 지금은 초기 치매환자들을 위해 자원봉사하고 있다고 한다.

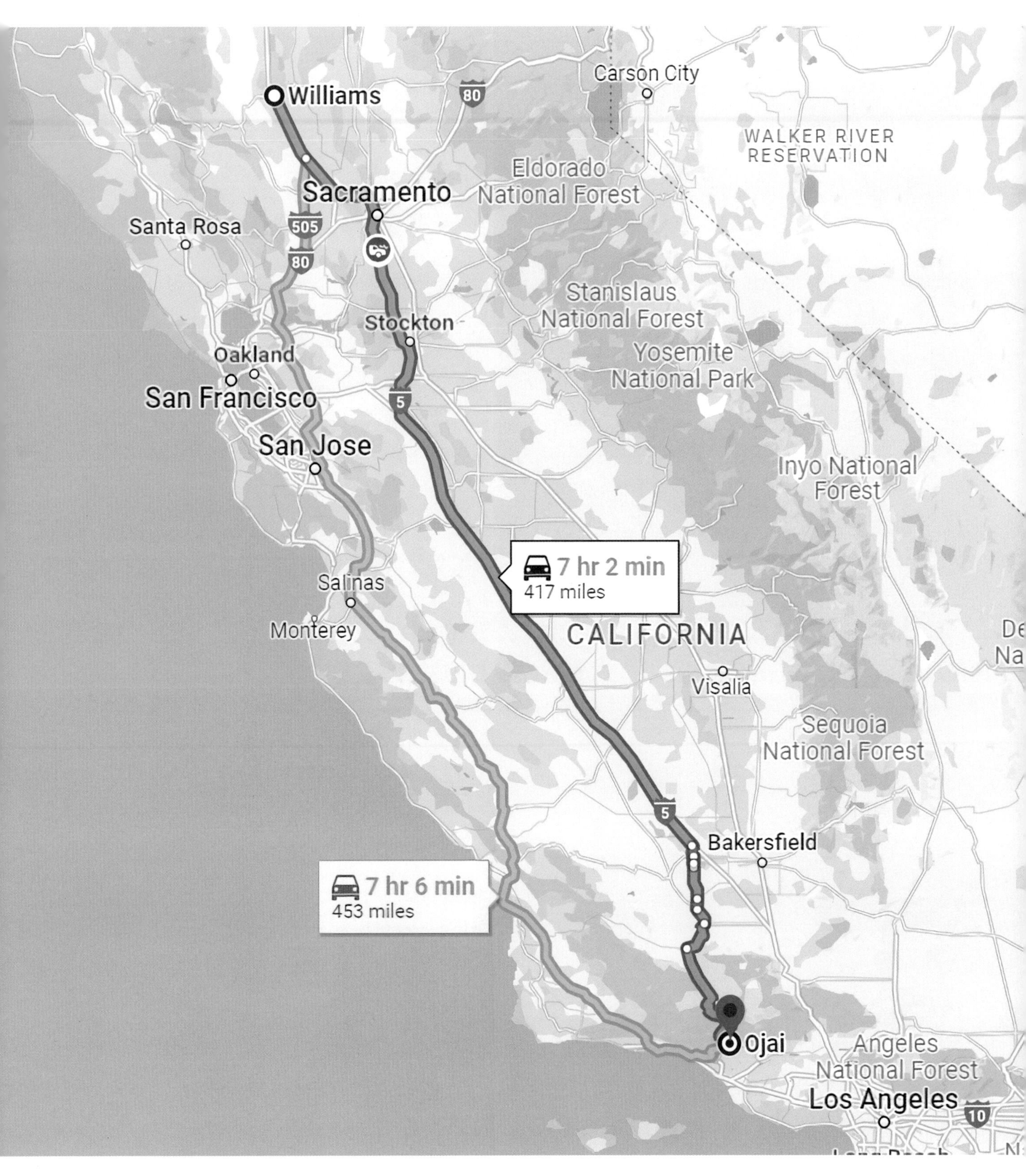

Williams → Ojai
417 miles (671 km)

윌리엄스 → 오하이, 671 km

Rocky, Carl, June and Jo.

록키, 카알, 준, 그리고 조

June and I had a breaktime
at a rest area.

휴게소에서…

At Jo's alma mater

조의 모교에서…

26th day

July 26th, Tuesday

26일째, 7월 26일 화요일

It is the 12th day of our return.

We're entering L.A. today.

Arrived in downtown L.A..

I'm so happy to see signs in Korean again.

Korean restaurants, Korean bathhouses, Korean barbershops, etc.

After first taking a warm bath, I plan to go to my favorite restaurant, Bukchang-dong Sundubu, for lunch, and then drop in a Korean bookstore.

We checked in at the New Seoul Hotel.

되돌아가기 12일째.

오늘 엘에이 입성.

드디어 엘에이 시내에 도착.

한국 식당, 한국 목욕탕, 한국 이발관 등….

한글 간판들이 너무나 반갑다.

목욕 먼저 하고는 단골집인 북창동순두부 식당에 가서 점심을 먹고

한국 서점을 둘러보았다.

뉴서울호텔 숙박.

Ojai → Los Angeles
82 miles (133 km)

오하이 → 엘에이, 133 km

Lunch at Bukchang-dong Sundubu.

점심 메뉴. 북창동순두부

Dried pollack hangover soup for dinner.

저녁 메뉴. 얼큰한 황태해장국

A piece of art that I like, hanging at the New Seoul Hotel.

뉴서울호텔에 걸려 있는 마음에 드는 작품

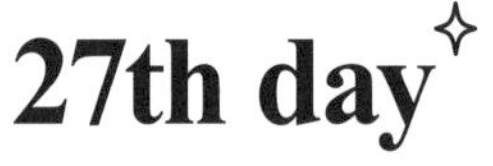

27th day

July 27th, Wednesday  27일째, 7월 27일 수요일

Today, we are having a reunion party with myself and 4 high school friends living in the L.A. area to celebrate the fact that I'm still alive after my long trip.

The wife of my friend Jong-don, prepared delicious bibimbap for us at their home. His wife usually prepares delicious things such as kimchi and side dishes, which I'm very grateful for. This was the first home-cooked Korean food that I'd had in a month, and it was very delicious. During the meal, for the first time in quite a while, we all had the chance to chat and have a jolly time.

On the way back home, I stopped by a McDonald's to have some vanilla ice cream with June.

We arrived home around 6 p.m.

Armando, Sonya(my tenants) and Kay(my Chihuahua) welcomed me. Thank God!

오늘 엘에이 근교에 사는 고교동창들 4명이 내가 살아 돌아온 기념으로 파티를 해줬다.

친구 종돈이의 부인이 집에서 비빔밥을 맛있게 준비해 줬다. 평소에 김치랑 밑반찬 등 맛있는 것을 챙겨주는 고마운 친구 부인이다. 한 달 만에 먹어보는 홈 메이드 한식, 정말 맛있었다. 식사하면서 수다도 떨고 오랜만에 즐거운 시간을 가졌다. 집으로 가는 길에 맥도날드에 들려 준과 바닐라 아이스크림을 먹으며 휴식했다.

오후 6시경 집에 도착.

내 세입자인 알만도, 쏘냐와 케이(나의 치와와)가 나를 반긴다.

감사합니다, 하느님!!

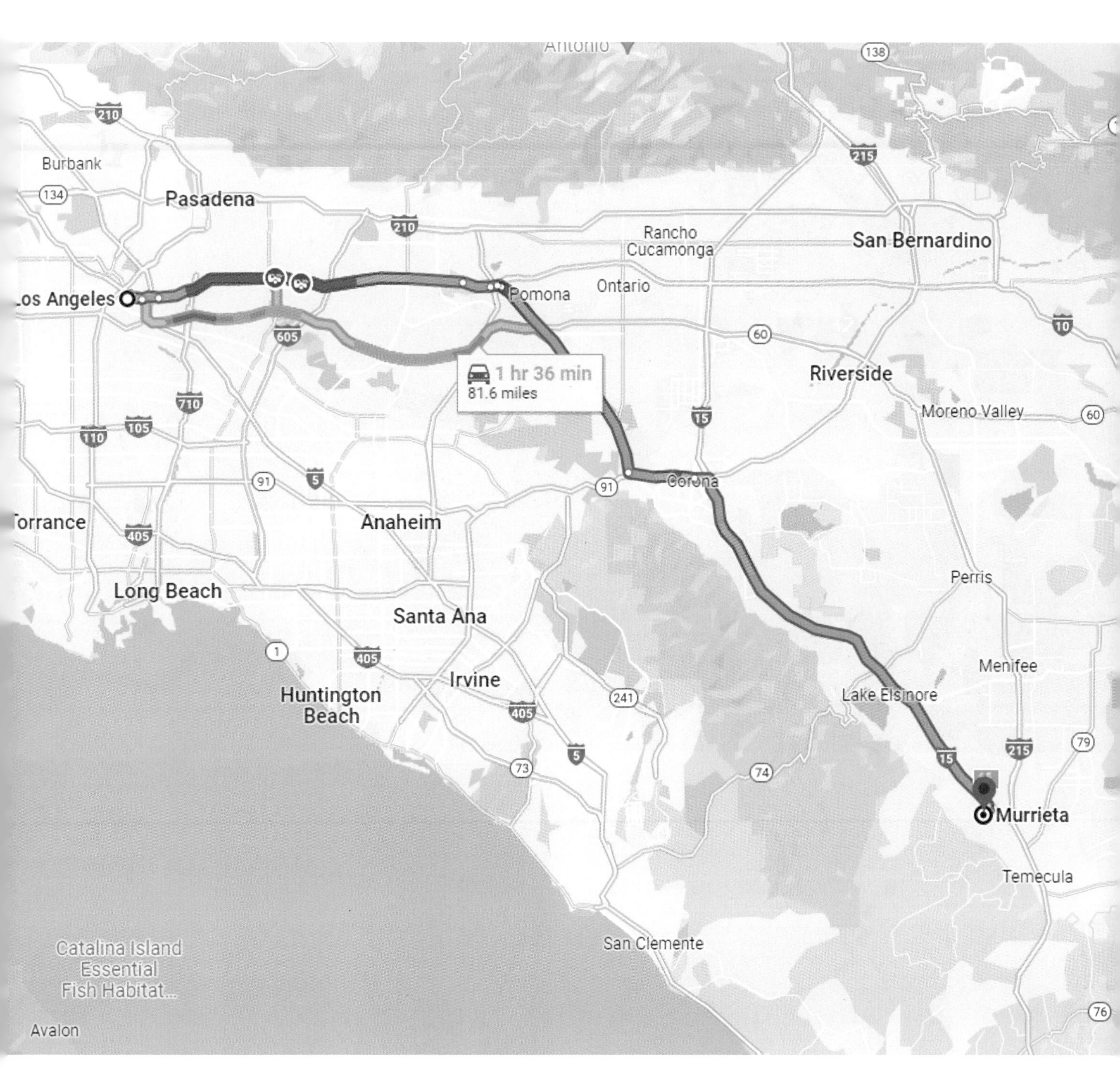

Los Angeles → Murrietal
81 miles (131 km)

엘에이 → 뮤리에타, 131 km

↑

Cho Jun, Kwak Chang-ryeol,
Ko Myeong-seong, Lee Jong-don,
and Ahn Kwang-yong.

종돈네집에서.
조준, 곽창렬, 고명성, 이종돈, 안광용.

→

June likes vanilla ice cream.

준은 바닐라아이스크림을 좋아한다.

Completing the next-to-last item on my bucket list … June and Carl have arrived safely at the Murrieta house!

Departing on July 1st, 2022, we completed the 7,986 mile (12,777 kilometer) round trip in 27 days.

Crossing the borders between California, Oregon, and Washington, U.S.A., and again passing the borders through British Columbia and the Yukon in Canada, culminating in a successful round-trip to Homer, the southernmost town in Alaska, U.S.A.!

Thank you for your support!

I am scheduled to return home to Korea on August 5th.

Thank God again!

나의 버킷리스트 중에 끝에서 두 번째 것을 무사히 완성했다. 준, 카알 무사히 뮤리에타 집에 도착!

2022년 7월 1일 떠난 지 27일만에 왕복 7,986 마일(12,777km)를 완주.

미국 캘리포니아주, 오레곤주, 워싱턴주 국경 넘어 캐나다 브리티쉬 콜럼비아주와 유콘 자치주를 지나 다시 국경 통과. 미국 땅 알래스카주 제일 남쪽 끝 마을 호머까지 왕복 자동차 여행을 성공적으로 마치다!

이제 8월 5일에 한국으로 귀국 예정!

여행은 떠나서도 기쁘고 돌아와서도 기쁘다.

여행의 처음과 끝 그리고 중간… 모든 순간에 감사드립니다!

하느님 감사합니다.

After the trip, I have often been asked what impressed me the most.
It is difficult to reply to that with a short answer.
The world is wide; it is packed with exciting things, and there are many things to see.
My heart beats faster, knowing the fact that all these things are waiting for me.

I was able to make eye contact with many bears and wild animals such as buffalo while driving on the highway.
The scene where a girl, who looked as if she was a middle school student, rides her horse to work, where she cleans my friend Larry's house.
Delightful Lillianne, who travels the roads in her Toyota passenger car with a summer house attached to its roof.
Amazing people who ride bicycles or walk across North America.
The Korean couple who run a grocery store deep in the Rocky Mountains, whom I was happy to meet.
A Korean couple at a Korean restaurant who served delicious Korean food in Homer, a small neighborhood at the southern tip of Alaska.
The time I drowsily drove, and entered the wrong road against opposing traffic by mistake and almost got killed …
They have all become precious and beautiful memories that cannot be omitted.
I offer you all my thanks, support, and love.

— Carl

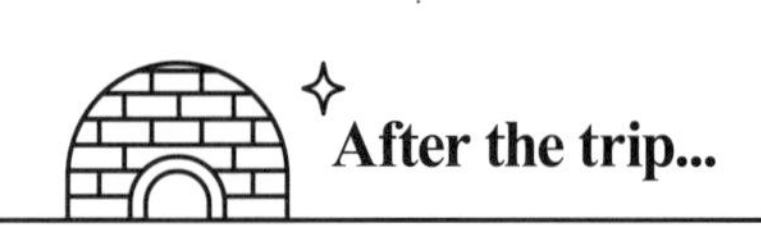

여행이 끝난 후, 제일 인상깊었던 것이 무엇이었냐는 질문을 많이 받는다.
단답형으로 대답하기가 어렵다.
세상은 넓고, 볼 것은 많고, 신나는 것들로 가득하다.
그런 모든 것들이 나를 기다리고 있다는 사실에 가슴이 뛴다.

고속도로를 달리면서 만난 많은 곰들과 버팔로(들소) 등 야생동물들과의 눈맞춤,
중학생으로 보이는 소녀가 자신의 말을 타고 출근해서 내 친구 래리 집을 아르바이트로 청소하는 장면,
도요타 승용차에 여름 별장을 싣고 다니며 여행하는 유쾌한 릴리안,
자전거를 타고, 혹은 걸어서 북미대륙을 횡단하는 대단한 사람들,
록키 산맥 산중에서 식료품점을 운영하는 반가웠던 한국부부,
알래스카 남쪽 끝 조그만 동네 호머에서 맛있는 음식을 내주던 한국식당의 한국부부…
두 번이나 실수로 도로를 역주행했고, 두 번의 졸음운전으로 자칫 죽을 뻔 했던 순간들…
어느 하나 빼놓을 수 없는 소중하고 아름다운 추억이 되었다.
모두에게 감사와 응원, 사랑을 보낸다.

— 2022년 10월 1일 토요일, 안광용

※All sales generated from this book will be donated
to the KKotdongnae Retreat in Temecula, CA. run by Father Oh Woong-jin,
and to the NAUH a North Korean human rights organization
founded by Rep. Ji Sung-ho.

※본 책의 판매금 전액은 오웅진 신부가 운영하는
테미큘라의 꽃동네와
지성호 의원이 개설한 북한인권단체 NAUH(나우)로
기부됩니다.

Carl & June's Alaskan Road Trip

Carl & June의 Alaska 자동차 여행일기

발행일	2022년 10월 8일
지은이	안광용
발행인	안광용
발행처	(주)진명출판사
교정교열	김영애, 김영신
등록	제10-959호(1994년 4월 4일)
주소	서울시 마포구 양화로 156, 1517호(동교동, LG팰리스빌딩)
전화	02-3143-1336
팩스	02-3143-1053
이메일	jmtax@jinmyong.com
정가	15,000원 (USA $15, Canada $20)

ISBN 978-89-8010-498-7 03980

착한 영어 시리즈
Pure and Simple English Series

※ 문의 | 010-6390-0945 김종규
010-5674-1336 김영애
010-4425-1012 안광용

※ 전화주시면
교사용 견본 및 MP3를
보내드리겠습니다.

01.
착한 여행영어
(포켓판)

02.
착한 영어 파파라치!
(틀린 영어 간판 찾기)

03.
착한 1,2,3 쉬운 생활영어
(한 단어, 두 단어, 세 단어로 된)

04.
착한 여행영어 일기
(남미편)

05.
착한 영문법
(기초 · 중급용)

06.
착한 생활영어
(미국 일상에서)

부록
핵심 영어문장 59개
자주 쓰이는 불규칙동사표
Konglish(한국에만 존재하는 영어)

07.
착한 여행영어 회화
(교재용)

08.
착한 기초영어 첫걸음
(왕초보를 졸업한)

09.
착한 미국영어 회화
(중 · 고급용)

10.
착한 업그레이드 된 기초영어 회화 첫걸음
(초 · 중급용)

11.
착한 서비스영어(비즈니스 영어)
(중 · 고급용)

12.
착한 팝송으로 배우는 영어회화

13.
착한 왕초보 영어회화 (근간)

(주)진명출판사
Tel. 02-3143-1336, 02-3143-1337 E-mail book@jinmyong.com

틀린 영어 간판을 찾아서
영어 공부를!

정가 10,000원

한 단어, 두 단어, 세 단어로 구성된 쉬운 생활영어!

정가 10,000원

무전 여행기를 통한
영어 독해·작문 향상시키기!

한 권으로 완벽히 끝내는 기초 문법!

정가 13,000원

미국 일상생활에서
많이 사용하는
현장 영어를 알고 싶다면!

정가 15,000원

독학 · 학원 강의용으로 만든,
매 챕터마다
연습문제와 복습 문제가!

정가 15,000원

초보분들, 초보의 단계를 지나신 분들, 학교를 졸업한 지 몇 십년 지나신 분들을 위한!

정가 15,000원

미국 현지인들이 즐겨 사용하는 일상언어와 문화!

Jokes! Curses! Idioms! Slang! Culture!

당신의 기초회화를 한 단계 Upgrade 시키세요!

초급 비즈니스(서비스) 영어의 일반적인 표현들부터 모든 업종에 필요한 표현들!

숙박업, 항공사, 운송업, 판매업, 금융업, 보안업, 미용업 등

착한 영어시리즈 11

정가 15,000원

착한 왕초보 영어회화

(근간)

(주)진명출판사